Hole-Drilling Method
for Measuring Residual Stresses

Synthesis SEM Lectures on Experimental Mechanics

Editor
Kristin B. Zimmerman, *SEM*

Hole-Drilling Method for Measuring Residual Stresses
Gary S. Schajer and Philip S. Whitehead
2017

Hole-Drilling Method for Measuring Residual Stresses

Gary S. Schajer and Philip S. Whitehead

ISBN: 978-3-031-79712-5   paperback
ISBN: 978-3-031-79713-2   ebook
ISBN: 978-3-031-79714-9   hardcover

DOI 10.1007/978-3-031-79713-2

A Publication in the Springer series

SYNTHESIS SEM LECTURES ON EXPERIMENTAL MECHANICS

Lecture #1
Series Editor: Kristin B. Zimmerman, *SEM*
Series ISSN
ISSN pending.

Hole-Drilling Method
for Measuring Residual Stresses

Gary S. Schajer
University of British Columbia, Canada

Philip S. Whitehead
Stresscraft Ltd., UK

SYNTHESIS SEM LECTURES ON EXPERIMENTAL MECHANICS #1

ABSTRACT

This book describes the theory and practice of the Hole-Drilling Method for measuring residual stresses in engineering components. Such measurements are important because residual stresses have a "hidden" character because they exist locked-in within a material, independent of any external load. These stresses are typically created during component manufacture, for example, during welding, casting, or forming. Because of their hidden nature, residual stresses are difficult to measure and consequently are often ignored. However, they directly add to loading stresses and can cause catastrophic failure if not properly included during engineering design. Thus, there is an urgent need to be able to identify and measure residual stresses conveniently and reliably.

The Hole-Drilling Method provides an adaptable and well-proven method for measuring residual stresses in a wide range of materials and component types. It is convenient to use and gives reliable results. Because of the hidden nature of residual stresses, the measurement method must necessarily be indirect, thus, additional care and conceptual understanding are necessary to achieve successful results. This book provides a practical introduction to the Hole-Drilling Method, starting from its historical roots and going on to focus on its modern practice. The various chapters describe the nature of residual stresses, the principle of hole-drilling measurements, procedures and guidance on how to make successful measurements, and effective mathematical procedures for stress computation and analysis. The book is intended for practitioners who need to make residual stress measurements either occasionally or routinely, for practicing engineers, for researchers, and for graduate engineering and science students.

KEYWORDS

hole-drilling, residual stresses, stress measurement, strain gauges, optical metrology, inverse methods

This book is dedicated
with respect and appreciation to

Josef Mathar

Originator of the Hole-Drilling Method

The research group of Theodore Von Kármán at the Technical University of Aachen. Josef Mathar was a member of that group and is likely among those in the photo. Unfortunately, because of wartime damage, no individual photograph of Mathar survives in the RWTH Aachen University archive. (Photo courtesy of the von Kármán Archive at the California Institute of Technology.)

JOSEF MATHAR – BIOGRAPHY

Karl August Martin Josef Mathar was born on 22 May 1900 in Magdeburg, Germany. He was educated at the high school in Euskirchen from 1911 to 1919. He went on to study at the Institute of Mechanical Technology and Machine Elements in Aachen, where he completed his Diplomhauptprüfung (qualifying exam) in December 1924. In 1925, he commenced work under the supervision of Prof. Dr. Ing. Rötscher. He continued as an Assistant at the Institute for Mechanics in Aachen in 1926, where he wrote a number of academic papers related to aircraft construction. At this institute Mathar wrote his thesis "On the Stress Distribution in Rod Ends" for which he was awarded the Dr.-Ing. degree in 1928.

In 1930 Josef Mathar defended his habilitation (faculty certification) thesis on the topic "Strength of Materials, with Specialization in Aircraft Construction" and became a Privat-dozent (Assistant Professor) in Aachen. He worked in the research group of Theodore von Kármán, who is regarded as the outstanding aerodynamic theoretician of the twentieth century. It was within this group that Mathar did his pioneering work on the development of the Hole-Drilling Method. He was also active in teaching and served as von Kármán's representative in lecture courses on aircraft statics.

Tragically, Josef Mathar died young at the age of 33 on 25 July 1933; his pioneering 1934 paper on hole-drilling was published posthumously. The significance and quality of Mathar's work was evidently well recognized at the time. Theodore von Kármán personally translated his paper into English.

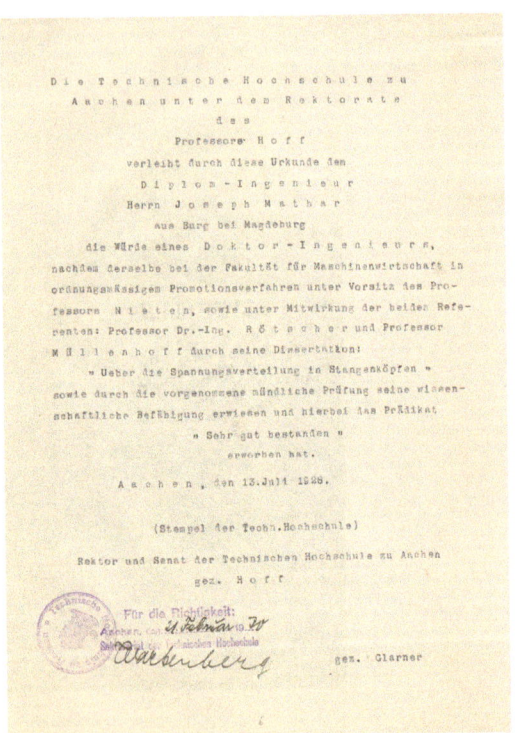

Transcript of Josef Mathar's Dr.-Ing. certificate. (Courtesy RWTH Aachen University Archive.)

Contents

Preface xiii

1   Nature and Source of Residual Stresses 1

   1.1   Introduction 1

   1.2   Origin of Residual Stresses 2

   1.3   Sources of Residual Stresses 4

      1.3.1   Bulk Component Misfit in Redundant Structures 4

      1.3.2   Non-Uniform Dimensional Variations due to Thermal Effects 6

      1.3.3   Non-Uniform Plastic Deformation 8

      1.3.4   Surface Machining 10

      1.3.5   Surface Treatments 11

      1.3.6   Chemical and Phase Change 12

   1.4   Types of Residual Stresses 13

      1.4.1   Residual Stress Type I 13

      1.4.2   Residual Stress Type II 13

      1.4.3   Residual Stress Type III 13

   1.5   Effects of Residual Stress 14

   1.6   Residual Stress Measurements 16

   1.7   Further Reading 17

2   Relaxation Type Residual Stress Measurement Methods 19

   2.1   Introduction 19

   2.2   Relaxation Method Concept 20

   2.3   Excision Method 21

   2.4   Two-Groove Method 22

   2.5   Splitting Method 24

   2.6   Slitting (Crack Compliance) Method 25

   2.7   Ring-Core Method 29

   2.8   Hole-Drilling Method 30

   2.9   Deep-Hole Method 32

   2.10 Layer-Removal Method 34

	2.11	Contour Method	35
	2.12	Sectioning Method	36
	2.13	Impact of Modern Measurement Technologies	37
	2.14	Method Selection	38
	2.15	Further Reading	41
3		**Hole-Drilling Method Concept and Development**	**47**
	3.1	Introduction	47
	3.2	Concept	47
	3.3	Mathar's Foundational Work	49
	3.4	Hole Drilling	49
	3.5	Deformation Measurements	53
	3.6	Ring-Core Method	57
	3.7	Deep-Hole Drilling	58
	3.8	Residual Stress Computations	59
	3.9	Concluding Remarks	62
	3.10	Further Reading	63
4		**Strain Gauge Technique: Method Description**	**69**
	4.1	Strain Gauge Rosette Selection	70
	4.2	Specimen Preparation	71
	4.3	Gauge Installation	73
	4.4	Instrumentation and Electrical Connections	76
	4.5	Hole-Drilling Equipment	77
	4.6	Hole-Drilling Procedure	80
	4.7	Gauge Data	85
	4.8	Further Reading	85
5		**Stress Computations**	**87**
	5.1	Introduction	87
	5.2	Uniform Residual Stresses	87
	5.3	Calibration Constants	91
	5.4	Stress Averaging	94
	5.5	Non-Uniform Residual Stresses	96
	5.6	Practical Determination of $\bar{a}$ and $\bar{b}$	100
	5.7	Regularization	103

5.8   Other Calculations 107
     5.8.1   Differential Strain and Average Stress Methods 107
     5.8.2   Power Series Method 108
     5.8.3   Specimen Thickness 108
     5.8.4   Hole Eccentricity Correction 110
     5.8.5   Plasticity Effects 110
     5.8.6   Orthotropic Materials 111
5.9   Further Reading 113

6   Example Practical Procedures and Results **119**
6.1   Specimen Geometry and Strain Gauge Selection Details 119
6.2   Practical Strain Gauge Rosette Installations 122
6.3   Orientation of Type B Strain Gauge Rosettes 128
6.4   Installation on Irregular Surfaces: Bond Thickness 129
6.5   Non-Standard Gauges 130
6.6   Residual Stress Example: Training Sample (Annealed Disc) 132
6.7   Residual Stress Example: Aluminium Alloy Block 133
6.8   Residual Stress Example: Machined, Forged Disc 135
6.9   Residual Stress Example: Surface Process Samples 135
6.10 Residual Stress Example: Thin, Shot-Peened Beam 137
6.11 Concluding Remarks 139
6.12 Further Reading 139

7   Optical Techniques **143**
7.1   Introduction 143
7.2   Holographic Interferometry 144
7.3   Moiré Interferometry 146
7.4   Electronic Speckle Pattern Interferometry (ESPI) 148
7.5   Digital Image Correlation 150
7.6   Computation of Uniform Residual Stresses 152
7.7   Computation of Non-Uniform Residual Stresses 158
7.8   Residual Stress Computation Using Incremental Data 160
7.9   Concluding Remarks 163
7.10 Further Reading 164

Authors' Biographies **167**

Index **169**

Preface

The Hole-Drilling Method is the most widely used relaxation type method for measuring residual stresses. Since its inception in the 1930s, the technique has dramatically grown and developed, with important new advances continually reported. The size and continued vitality of the research literature on the subject testifies to the fertility and rich potential of the technique. Josef Mathar, to whom this book is respectfully dedicated, would have had good reason to be proud of the evolution of his elegant concept. It is very unfortunate that his life was cut short so that he did not live to see the early flowerings of his foundational work.

While it may be said that the Hole-Drilling Method has one ancestor, it also has very many family members, each of whom brings an important and diverse contribution to the technique. Josef Mathar certainly laid down the foundation of the method, in his published work he also conceptually anticipated both the Ring-Core and Deep-Hole methods. However, the great strength of the Hole-Drilling Method derives from the numerous subsequent workers who substantially and very skillfully expanded the concept. It was they who made it into the robust and practical technique that it is today.

The authors are among the hole-drilling "family members," each with experience of the method extending over many years. It is our wish to share this experience with interested readers and thereby contribute to the wider use of the Hole-Drilling Method and to its ongoing development. This book is primarily intended as a practical handbook, although it is also hoped that it will find good secondary use as a reference textbook. The target audience is measurement practitioners, practicing engineers and students. In support of their needs we have endeavoured to make the book contents as accessible as possible, while seeking to maintain a rigorous technical standard. In this respect, our objectives parallel the idea indicated the title of a totally unrelated work, *The Prepared Table*, a classical book of religious law written in the 16th century by Joseph Karo. Neither of us have read the very specialized content, but the mental image conjured by the inspired title vividly demonstrates the need to present a written work in a well-laid out and accessible way analogous to how one would arrange the food on a dinner table for a welcome guest. It is the task of the host to prepare the table so that the guest will find well-made and satisfying things ready for them. We have endeavoured to achieve this objective within this book. How well we may have succeeded will be up to the reader to judge. In places where we have fallen short, it is hoped that a generous allowance may be granted.

The substantial size and variety of the hole-drilling literature has made it very challenging for us to arrange our prepared table. There are so many "foods" that we could serve, but excess provision brings serious risk of indigestion. Thus, we have tried to follow a "keep it simple" approach where we focus on the basic information. Each chapter also includes a thematic list of further readings so the reader can find further details on particular topics of interest. These readings are ones with which we happen to be familiar, but they certainly are not the only suitable materials, possibly not even the best. Certainly, there is no adverse opinion implied by the non-inclusion of other important work; for sure there is no monopoly of wisdom.

No enterprise can be undertaken without the substantial help of others. Our families, friends and colleagues have generously helped and supported us in the preparation of this book. Our parents have educated us and enabled us to reach our potential; this book is a flowering of their efforts and is offered as a sign of our gratitude. Special thanks are also offered to the staff members at Stresscraft Ltd., who all helped to prepare the strain gauge rosette installations and drilled the holes for the examples included in this book; they are: Jeremy Lodge, Calum Pratt and Patricia, Charlotte, Ian, Jamie and Matthew who are all members of the "Whitehead" family. Also thanks to Prof. Philip Withers, Dr. Michael Prime, Prof. Drew Nelson, Prof. Armando Albertazzi and Dr. Matias Viotti for their skillful reviews of the chapters. Prof. Lyndon Edwards gave the idea of including quotes at the start of each chapter. Finally, Mr. Juuso Heikkinen proof read the entire manuscript, suggested many clarifications, and rooted out numerous typos. Thank you all!

In keeping with the "pay forward" principle, all author royalties from the sale and distribution of this book are contributed to the Leonard and Lilly Schajer Memorial Bursary Fund at the University of British Columbia, Canada. Proceeds from the fund are used to support Mechanical Engineering students in financial need. From among such students will come our next generation of residual stress practitioners. We hope that we have prepared the table well for them and we wish them every success.

Gary S. Schajer and Philip S. Whitehead
December 2017

CHAPTER 1

Nature and Source of Residual Stresses

"No materials, components or structures are completely free from residual stress."

Viktor Hauk (1997) in *Structural and Residual Stress Analysis by Non-Destructive Methods.*

1.1 INTRODUCTION

Residual stresses are "locked-in" stresses that exist within a material without the presence of any external loads. They are "self-equilibrating," that is, there exist within the material localized areas of tensile and compressive stresses such that of the effects of the tensile areas balance those of the compressive areas to give zero force and moment resultants. Figure 1.1 illustrates the residual stresses within a block of heat-tempered glass. To improve its strength, the material is manufactured such that its surfaces are in compression, balanced by tension in the interior. Figure 1.1a shows the distribution of the horizontal normal stresses along a central vertical plane and Figure 1.1b shows a photoelastic fringe pattern measured through the material depth.

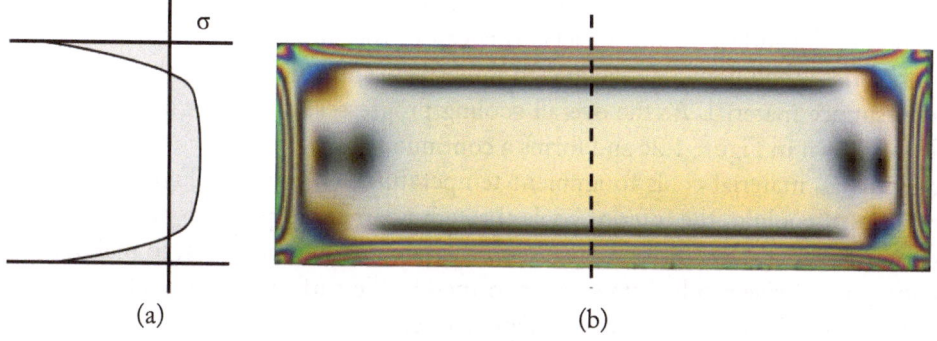

(a)   (b)

Figure 1.1: Stress distribution in thermally toughened glass. (a) Horizontal stress profile on a central vertical plane and (b) photoelastic stress measurement (adapted from Scafidi et al. (2015)).

For horizontal equilibrium with zero force resultant, the area of the tensile (+) stress region shaded in Figure 1.1a equals the sum of the areas of the compressive (−) stress regions. The presence of the compressive stress regions also at the left and right sides of the photoelastic measurement in Figure 1.1b illustrates that equilibrium is obeyed on all planes, giving zero force resultants both horizontally and vertically. Similar zero in-plane force resultants occur with the shear stresses. In this example, moment equilibrium is automatically obeyed because of the stress symmetry, but it is also obeyed for the general non-symmetric case.

Because they exist locked-in within a material that has no external loads, residual stresses seemingly have a "hidden" character. They give few readily apparent indications of their presence and so can easily be overlooked. There is also sometimes a tendency among designers to shy away from considering residual stresses because they can be difficult to measure, predict or analyze. However, inclusion of residual stresses is very important for practical applications because they add to the structural stresses due to the applied loads and can possibly make the total stress much different than anticipated. Inclusion of residual stresses in engineering design analysis has always been an important need, but it is increasingly urgent as modern structures become lighter and less conservative.

1.2   ORIGIN OF RESIDUAL STRESSES

The combination of tensile and compressive stress regions within a material that characterize residual stresses are caused by "misfits" among those regions. For example, the stresses shown for the heat-tempered glass illustrated in Figure 1.1 occur as a result of dimensional differences between the outer surface material and the interior material. During manufacture, the glass is heated to become a liquid, formed into a flat shape and then cooled to become a solid glass sheet. Figure 1.2 schematically illustrates the steps in the process. Figure 1.2a shows the glass as a uniformly very hot liquid. For clarity, the cross-section is divided into three conceptual regions comprising the two surfaces and the interior. The solidification step in Figure 1.2b is done by rapidly cooling the surfaces using air jets. This causes the surfaces to solidify while the hotter interior material is still liquid, allowing it to flow freely and adjust to the thermal shrinkage of the cooler surface material. As the overall cooling progresses, the hotter interior material also solidifies as shown in Figure 1.2c and forms a continuous solid with the cooler surface material. Subsequently, all material cools to ambient temperature, as shown in Figure 1.2d. However, during this final cooling, the interior cools through a larger temperature range than the outer surfaces. Consequently, it shrinks more due to thermal contraction. Thus, to maintain material continuity, compressive residual stresses are formed in the surfaces, balanced by tensile stresses in the interior, as schematically shown in Figure 1.2e.

The thermally toughened glass example of Figures 1.1 and 1.2 illustrates a case where residual stresses are deliberately created. However, material misfits and their associated residual stresses are induced by almost all manufacturing methods, e.g., casting, grinding, forming and welding. Indeed, the name residual stresses derives from their creation as a residue from such

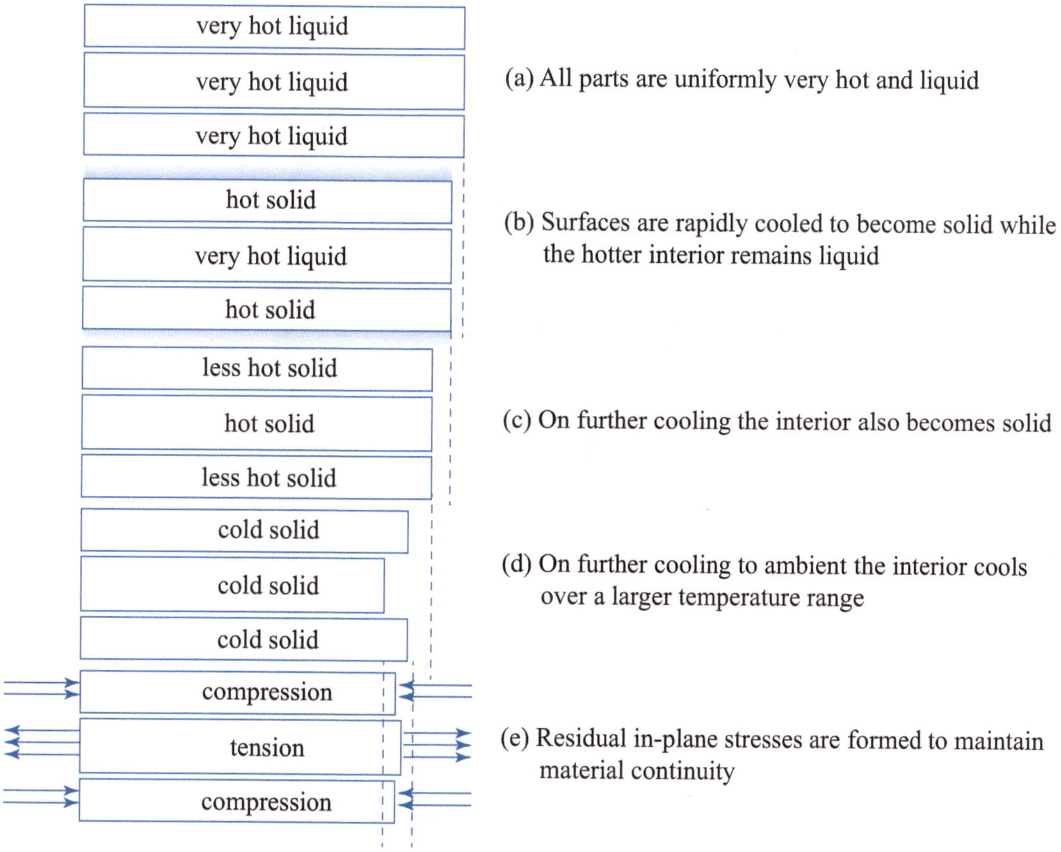

very hot liquid	
very hot liquid	(a) All parts are uniformly very hot and liquid
very hot liquid	
hot solid	
very hot liquid	(b) Surfaces are rapidly cooled to become solid while the hotter interior remains liquid
hot solid	
less hot solid	
hot solid	(c) On further cooling the interior also becomes solid
less hot solid	
cold solid	
cold solid	(d) On further cooling to ambient the interior cools over a larger temperature range
cold solid	
compression	
tension	(e) Residual in-plane stresses are formed to maintain material continuity
compression	

Figure 1.2: Residual stress creation in thermally toughened glass.

prior processing. Figure 1.3 shows an example of residual stresses within an optical lens. In this example, insufficiently slow cooling during solidification of the lens material created significant residual stresses in the final product, as revealed by crossed isoclinic lines seen when viewing in plane polarized light. These lines show the locations where the principal stress directions correspond to the principal polarization directions. This method of observation in plane polarized light is a common quality control step in lens manufacture. The response shown in Figure 1.3 is very severe and would be unacceptable for a good quality lens.

The material misfits that are the source of residual stresses are called "inherent strains" or "eigenstrains." The measurement and identification of these strains provides a powerful method for quantifying residual stresses. An idealized "stress-free" material without any interior misfits could be a perfect single crystal of a metal or a perfectly uniform amorphous material created through a quasi-infinitesimal cooling rate. However, all practical materials have some amount

Figure 1.3: Lens stresses revealed by observation in plane polarized light.

of structural or processing non-uniformity, for example the grains within a metal or the phase changes within a hardened surface. Thus, all practical materials have some level of residual stress.

1.3   SOURCES OF RESIDUAL STRESSES

Residual stresses can be created from many different sources, not just by the non-uniform so-lidification mechanism described in Figure 1.2. Any process that causes misfits among different parts of a material or structure will induce residual stresses. Most sources of residual stress can be described in terms of the following four general categories.

1.3.1   BULK COMPONENT MISFIT IN REDUNDANT STRUCTURES

Redundant structures are structures that have more than the minimum number of supports or members required to bear the applied loads. Most practical structures are redundant to some extent, some quite substantially so. Such redundancy can often be desirable because it enables a structure to continue to function even after one or more parts have failed.

A major characteristic of redundant structures is that misfit among the excess supports and members can create internal stresses in addition to those from external loads. These internal "fit-up" stresses are residual stresses that exist within the structural members as a result of their assembly into that structure. If the structure were disassembled, the associated residual stresses in each member would disappear. In addition to the failure tolerance that structural redundancy can provide, the associated residual stresses could also possibly enable the structure to function more effectively. In such cases the residual stresses may be deliberately created, for example in pre-stressed concrete, bolted and riveted connections, and shrink-fitted components.

Conversely, in many cases the residual stresses that exist in redundant structures can become large and damaging. Assembly misfits and support settlement commonly cause large residual stresses in structures such as bridges and buildings. Environmental temperature changes can similarly create substantial residual stresses. A well-known example of this effect are "sun-kinks," which can occur in continuously welded railway rails in hot weather. The longitudinal constraint imposed by the continuous welding together with the latent thermal expansion due to the hot weather creates very large compressive axial forces. If the ballast of the railbed does not provide sufficient lateral constraint, the compressive axial forces cause the rails to buckle, sometimes quite dramatically. Figure 1.4 shows some examples of rail buckling recorded by the US Department of Transportation. They reported an average of 38 train derailments per year in the United States due to track buckling during the period 1992–2002.

Figure 1.4: Thermal buckling of railway rails, also called "sun-kinks" (photo courtesy of Volpe, The National Transportation Systems Center, USDOT).

High structural redundancy increases the susceptibility of assembled structures to the development of residual stresses. Thus, engineering designs often include features to reduce structural constraint, such as gaps between railway rails (now a deprecated choice because of the resulting increase in vibration and noise), thermal expansion joints in pipelines and shear and pin supports in bridges. Figure 1.5 shows examples of constraint reduction features in large structures.

Figure 1.5: Constraint reduction features to reduce structural residual stresses. (a) U-bend in a hot-water distribution pipe, (b) shear supports in a concrete railway bridge, and (c) pinned joints in a steel road bridge.

1.3.2   NON-UNIFORM DIMENSIONAL VARIATIONS DUE TO THERMAL EFFECTS

Local dimensional misfits within a material are commonly created by thermal processes during the material manufacture. Non-uniform cooling from a molten state to a solid state, such as occurs in the production of thermally toughened glass, Figures 1.1 and 1.2, is an example. In this case, the temperature non-uniformity is deliberately introduced to create desired residual stresses. However, in many cases the stresses are undesirable, and need to be controlled, typically by using slow cooling rates or by holding at a high temperature to allow the material to equilibrate. If not well controlled, very large stresses of this kind can be developed in castings, leading to dimensional instability during subsequent use or, in extreme cases, cracking of the cast ingot, as illustrated in Figure 1.6.

Welding is also well-known to produce very large residual stresses, reaching up to the tensile yield stress longitudinally within the weld. These stresses produce the type of dimensional distortion schematically illustrated in Figure 1.7a. Analogous lateral effects produce the distortion illustrated in Figure 1.7b. These effects can be reduced by control of welding technique, fixturing and stress relief of the finished weld by post annealing, shot peening and vibratory stress relief.

The formation of residual stresses in quenched components has several similarities to the process described for the solidification of a casting. For a quenched component, the starting temperature for the process is below the fusion temperature but is sufficiently high such that the yield strength of the component material is significantly smaller than the strength at room temperature. As the component is immersed in the quenching medium, usually water, oil or air, the temperature of the component skin reduces rapidly and the yield strength of this material increases. As the quench proceeds, the thickness of the cooler skin increases, while the reduction

Figure 1.6: Cracking of a cast aluminum ingot due to excessive residual stresses (photo courtesy of Mark Newborn, Alcoa).

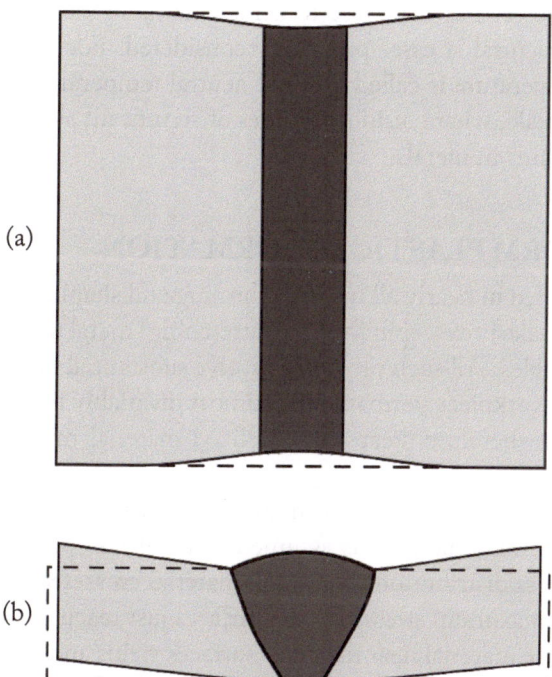

Figure 1.7: Geometrical distortions caused by welding residual stresses. (a) Plan view showing in-plane effects and (b) side view showing out-of-plane effects (adapted from Wizard191, Wikimedia Commons).

of the temperature at the core is much slower. As cooling proceeds, the imbalance of temperatures and associated dimensions between the core and skin results in generally tensile residual stresses at the core and balancing compressive stresses at the skin. Depending on the material, temperature ranges and quench medium, residual stresses can approach or exceed the material yield strength. In extreme circumstances, stresses during the quenching process can exceed the material yield strength or the ultimate strength resulting in fracture and partial relaxation of the residual stresses.

Composite materials bring into play a further mechanism for thermal stress development. In general, the individual materials within a composite structure have different physical properties. Thus, differences in thermal properties produce differential expansions among the various materials, causing dimensional misfits and consequent residual stresses. If large, the stresses can cause permanent deformations and associated residual stresses. Smaller residual stresses may also occur in the elastic range and vary approximately linearly with overall temperature. These latter stresses are temporary insofar as a temperature cycle that returns to the starting temperature produces residual stresses that similarly cycle and return to the starting stress state. In some cases, there may be a particular temperature, perhaps corresponding to the manufacture temperature, at which the residual stresses become close to zero. This same mechanism occurs on a much larger scale with the structural stresses previously considered. For thermal stresses in railway rails, the zero-stress temperature is called the "rail neutral temperature." Similar behaviors can occur on a microscopic scale, where residual stresses of significant size can occur in and around inclusions and crystal grains in metals.

1.3.3   NON-UNIFORM PLASTIC DEFORMATION

Residual stresses are created in nearly all material forming and shaping procedures, for example, bending sheet metal to make boxes, spin forming (stretching) metal to make bowls, and twisting wires to assemble into cables. All such processes involve substantial plastic deformation so as to change the shape of the workpiece permanently. Almost invariably the associated plastic strains are non-uniform, and so therefore create the localized material misfits that produce residual stresses.

Figure 1.8 illustrates residual stress formation in a beam that is bent beyond its elastic range. For simplicity, the material here is assumed to be elastic-perfectly plastic. Figure 1.8a shows the maximum stress distribution within the material cross-section that can occur in the linear elastic range. The maximum stresses at the surfaces just reach the yield stress. On further bending in Figure 1.8b the material near the outer surfaces yields and deforms plastically at constant stress to produce the outer flat areas in the stress profile. The inner linear area remains in the elastic range. If the beam is then unloaded, elastic stress release corresponding to the dashed line in Figure 1.8b occurs. A surface stress greater than yield is required to give a linear stress distribution with equal moment resultant to the plastic stress distribution. This larger elastic stress change can occur because the available elastic unloading range spans from tensile yield, through zero to compressive yield and thus is twice the size of the elastic loading range. Figure 1.8c shows the residual stress resultant that remains after the elastic unloading. It corresponds to the plastic stress profile in Figure 1.8b minus the linear elastic unloading stress profile.

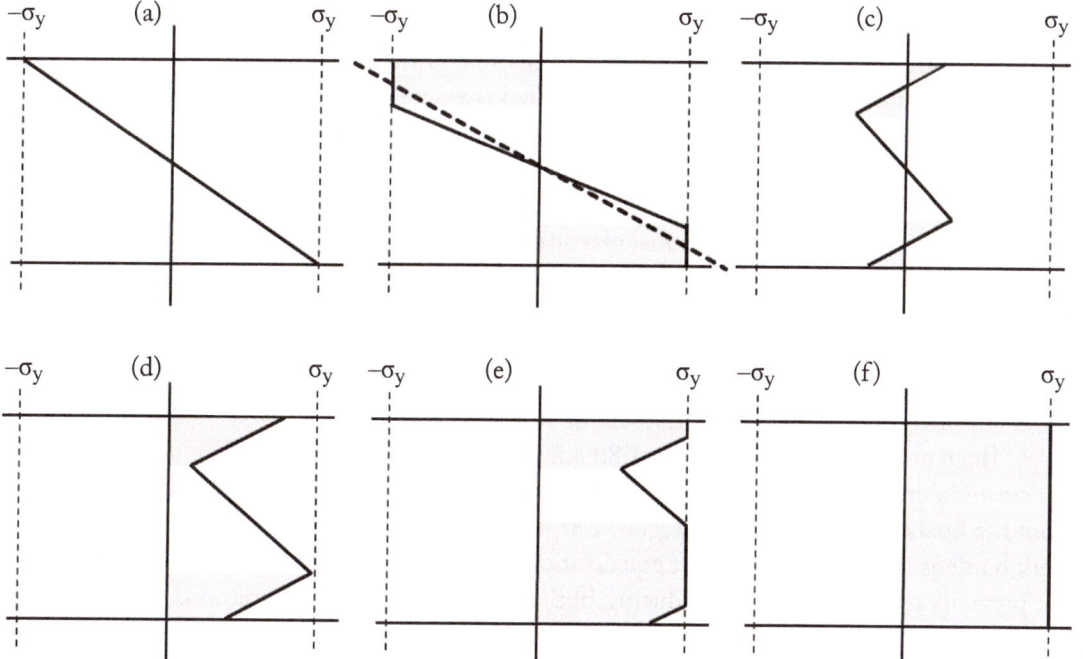

Figure 1.8: Residual stress creation and relief in a beam of elastic-perfectly plastic material. Shaded areas = plastic deformation. (a) Bending in elastic range, (b) bending into plastic range, dashed line = elastic moment resultant, (c) residual stresses after release of bending load, (d) addition of axial tension, (e) further addition of axial tension with plastic deformation, and (f) further addition of axial tension with full plastic deformation.

The bending process in Figure 1.8 illustrates a common feature of loading processes that create residual stresses. If the beam in Figure 1.8 were reloaded, it would reload elastically until it reached the plastic stress state previously reached during the prior loading in Figure 1.8b before further plastic deformation could occur. This is a larger applied load range than was initially available in Figure 1.8a. Thus, the residual stresses in Figure 1.8c are seen to have a strengthening effect. Similar behavior occurs with other loading types and such controlled overloading is often used as a means of strengthening against subsequent loads in the same direction. For example, cold hole expansion is commonly used to strengthen holes by controlled plastic overloading of the hole edge. Similarly, flywheels can be strengthened by initial running over-speed. The process of over-pressuring cylinders, called "autofrettage," is long-established as a method for strengthening cannons for military use. In all these examples, it should be noted that the strengthening effect is solely due to the presence of the residual stresses induced by the initial overload. The effect is separate from any strain-hardening that may also occur. Figure 1.8 shows that the strengthening effect occurs with a perfectly plastic material that does not strain harden at all.

Figures 1.8a,b,c show how local plastic deformation can create material misfits and resulting residual stresses. This is the most common situation. However, it can also happen that global plastic deformation can relieve misfits and residual stresses. If an axial tension were applied to the unloaded beam in Figure 1.8c, all stresses would uniformly become more tensile. Figure 1.8d shows the resultant stresses when the maximum stress just reaches the tensile yield stress. This loading is purely elastic. Figure 1.8e shows the stress distribution that occurs when the axial tension is increased such that additional plastic deformation occurs. As can be seen, the effect is to reduce the variation in the stress profile across the beam width. With yet further axial tensile load, as in Figure 1.8f, the plastic deformation spreads across the full height of the beam. All material is at the yield stress, thereby reducing the stress variation into a uniform profile. This response has effectively equalized the material misfits created by the initial bending. Thus, when the beam in Figure 1.8f is unloaded, it will be residual stress-free.

The tensile stretching in Figures 1.8d,e,f, called "stress leveling," does not reverse the misfits created during the initial bending in Figures 1.8a,b,c it "equalizes" them. After load release from the final tensile stretching in Figure 1.8f, the beam is longer than initially. A material that work hardens will not give a perfect equalization of the misfit distribution, but will "wash out" the previous residual stresses by reducing both their size and spatial variation. Figure 1.8e illustrates the reduction in spatial variation. Manufacturing processes such as wire drawing and extrusion involve substantial plastic deformation, but typically produce only modest residual stresses because of this "wash-out" effect.

1.3.4   SURFACE MACHINING

During component manufacture, many material removal processes—such as turning, milling and drilling—involve the application of highly localized mechanical and thermal loads. The newly formed surfaces typically contain significant residual stresses. Stress magnitude and penetration depth typically depend on the rate of material removal and the geometry/sharpness of the cutting tool, the material hardness and the use of lubricants/coolants. For many machining processes, the direct effects of machining may penetrate to depths within the range 50 μm ("fine" machining) to 1 mm ("rough" machining). Near-surface tensile stresses in the direction of machining are often greater than those in the transverse direction; details of tool geometry can also influence the magnitudes of shear stresses and the resulting principal stress directions.

In many grinding processes, the temperatures reached at the cutting surface usually exceed those for conventional machining, while mechanical loads may be somewhat smaller. Accordingly, thermal loading may be the predominant producer of residual stresses at the newly-formed surface. Because of the relatively low mechanical load input, near-surface residual stresses may be similar in the grinding and transverse directions. The use of un-cooled abrasive discs or "abusive" grinding may produce tensile residual stresses that extend to depths greater than 2 mm.

Material removal using low-speed abrasion, e.g., honing, can result in the creation of small, near-surface residual stresses that penetrate just a few microns. Such stresses are difficult to detect using relaxation methods such as hole-drilling and may require diffraction to quantify the resultant distributions.

Maximum temperatures generated during wire or die-sinking electro-discharge machining (EDM) generally exceed those attained during most grinding processes. As a result of the high process temperatures, molten material not ejected from the surface and removed by the flow of dielectric fluid re-solidifies at the surface to produce a characteristic "re-cast layer." Residual stresses in this layer are always tensile. Because of difficulty in the precise control of adequate flushing of dielectric fluid in die-sinking EDM, the resulting penetration depth of tensile residual stresses is often greater than for wire EDM.

1.3.5 SURFACE TREATMENTS

Because many material forming and removal process create tensile residual stresses at or close to the surface, methods have been developed for treating surfaces so that the tensile and often variable near-surface stresses are replaced by regular and predictable compressive stress. These methods include the following.

- **Shot-peening** is a process where small hardened balls, usually steel or iron, are blasted onto the material surface to cause local plastic deformation and induce surface compression. Depending on the size of the shot and intensity of the blasting process, the resulting compressive stresses can penetrate to depths in the range 100 μm–1.5 mm. For thin sections or delicate components, ferrous shot can be replaced by small glass spheres.

- **Laser shock-peening** produces layers of compressive stress that extend to depths significantly greater than those produced by shot-peening. For this process, an intense laser beam is directed at the target surface. The resulting rapid rate of heating at the surface produces an intense shock wave, generating in-plane compressive residual stresses at the surface. Compressive stresses to depths exceeding 2 mm can be achieved while exposing the component to lower levels of surface damage and contamination than shot-peening.

- **Deep rolling and roller burnishing** are methods of cold working component surfaces to generate compressive residual stresses and improve surface finish. In both cases a hardened roller or sphere is pressed against the target surface and moved in a controlled manner. Depending on the material and the roller geometry, compressive stress penetration in excess of 2 mm can be achieved. While cylindrical rollers can be readily applied to shafts and cylinder bores, burnishing using spherical rollers can be used on irregular surfaces and controlled in much the same way as cutter tool paths using CNC machines.

1.3.6 CHEMICAL AND PHASE CHANGE

This source of residual stress occurs when a chemical reaction and/or phase change takes place within part of a material whose products have different dimensions than the source, thereby creating a misfit and associated residual stresses. Phase changes commonly occur during heat treatment of metals. Additional thermally induced residual stress may also be created by the rapid quenching sometimes involved. In most cases, the resulting development of residual stresses is undesirable. However, occasionally material phase changes are deliberately introduced to form desired residual stresses. Figure 1.9 shows an example of a steel circular saw that has been "heat-tensioned" by localized induction heating near the tooth-line. Alternatively, oxy-acetylene or laser heating are also used in practice. The material in the heated area transforms from a martensite to a bainite structure and thereby undergoes a small reduction in volume and consequent circumferential tension. This tooth-line tension has the same beneficial effect on circular saw cutting stability as blade tightening has on a conventional hacksaw.

Figure 1.9: Circular sawblade heat-tensioned by induction heating.

These four sources of residual stress are just general categories. Residual stress examples could easily belong to more than one of them, depending on process details and which aspect of the residual stress production process is emphasized. For example, coated materials formed by electroplating, chemical vapor deposition, plasma deposition or glazing, typically have large residual stresses within the coated layer. The stresses in these coatings could be caused by phase-based effects during deposition, or by thermal expansion mismatch. These stresses can cause curvature in thin components; measurement of that curvature forms the basis of Stoney's method for plating stress evaluation.

1.4   TYPES OF RESIDUAL STRESSES

Withers (2001) defines three types of residual stress according to the scale over which they self-equilibrate. The distance over which self-equilibration occurs is called the characteristic length ℓ_0. This length influences how the residual stresses affect material properties and also the choice of potential stress measurement methods.

1.4.1   RESIDUAL STRESS TYPE I

These are macro scale stresses that vary quasi-continuously across dimensions comparable with the size of the component. The characteristic length ℓ_0 would commonly be measured in mm, but could be much larger for major structural stresses. Type I residual stresses are typically the stresses used by engineers for structural design purposes and are the objectives of most residual stress measurements.

1.4.2   RESIDUAL STRESS TYPE II

These are stresses on a microscopic scale, for example, on the scale of the grain structure of metals. Figure 1.10 schematically illustrates the Type II residual stresses that are created by inter-granular misfits. The crystal orientations within the grains are random and the elastic/plastic/thermal properties within each grain vary discontinuously from grain to grain. Here, the characteristic length ℓ_0 would span several grains and commonly be measured in microns. Type II residual stresses are of interest to materials scientists to understand material behavior and to develop improved materials and processing methods. High-cycle fatigue crack initiation is strongly influenced by both Type II and Type I residual stresses.

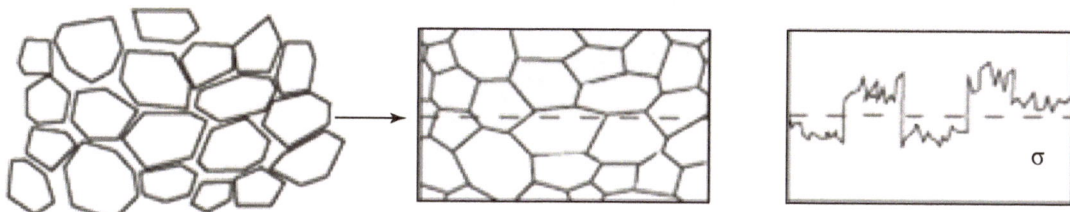

Figure 1.10: Type II stresses from intergranular misfits (adapted from Withers (2007)).

1.4.3   RESIDUAL STRESS TYPE III

These are stresses on an atomic scale, for example within a crystal lattice in the region of a dislocation, or at grain boundaries (coherency stresses). These irregularities in crystal structure produce highly localized variations in the force fields around the individual atoms within the crystal. The associated characteristic lengths ℓ_0 have the same scale as the lattice spacing and would commonly be measured in Ångström or nm.

1.5   EFFECTS OF RESIDUAL STRESS

Residual stresses can be beneficial or harmful depending on their character and distribution. Uncontrolled residual stresses that are large both in magnitude and in spatial extent are typically detrimental. Because of the large thermal gradients that occur during their manufacture, castings typically contain large residual stresses. In extreme cases, this causes the cracking illustrated in Figure 1.6. The geometry of cast metal components is commonly designed to ameliorate either the size or effect of residual stresses, typically by the inclusion of generous fillets to minimize stress concentrations and by the choice of favorable geometric features. Figure 1.11 illustrates a cast flywheel designed with curved spokes. At first glance, the spoke shape seems to be a decorative feature, but actually is deliberately designed to increase elastic flexibility within the wheel and thereby reduce the effects of the residual stresses that are inevitably present. Large fillets are also evident at the spoke connections.

Figure 1.11: Curved spokes used to reduce residual stresses in a cast iron flywheel.

Commonly, surface compressive stresses are beneficial because they close any surface cracks that are present and thereby increase the strength of brittle materials. The thermally toughened glass shown in Figure 1.1 is an example where surface compressive stresses are deliberately introduced to enhance material strength. Shot-peening is a further example of such material treatment. In that case, small hardened balls are blasted onto the material surface to cause local plastic deformation and induce surface compression. Pre-stressed concrete is used for broadly the same reason, where the steel reinforcing bars are deliberately tensioned so as to draw the surrounding brittle concrete into compression.

In contrast, tensile residual stresses are harmful in brittle materials and in materials prone to stress corrosion cracking. They are also harmful in materials subject to repeated loading because their adverse effects on fatigue life and/or allowable alternating stress amplitude. The Goodman and Gerber diagrams used for fatigue design highlight this fundamental effect. The associated loss of fatigue load capacity coupled with the "hidden" character of residual stresses can lead to unexpected structural failures, sometimes on a large scale and with very serious consequences. For example, residual stresses in a cast eyebar link were a major factor in the 1967 failure shown in Figure 1.12 of the U.S. 35 Silver Bridge in West Virginia. The collapse caused the loss of 46 lives.

Figure 1.12: Section of U.S. 35 Silver Bridge that collapsed in 1967 due to crack propagation associated with residual stresses (photo courtesy of U.S. Dept. of Transportation, Federal Highway Administration).

Wood is an unusual material because cell wall buckling causes it to behave oppositely to brittle materials and to have a compressive strength significantly lower than its tensile strength. In that case, the presence of tensile stresses at the surface can be advantageous, causing several species of trees to develop "growth stresses." These naturally occurring residual stresses comprise axial tension around the outside of the tree, balanced by axial compression in the interior. Wind-induced bending applies symmetrical tensile and compressive axial stresses that are largest at the outside of the tree. The presence of the growth stresses biases the surface stresses toward tension, thus mitigating the lower compressive strength. Figure 1.13 shows a eucalyptus tree after felling.

Disturbance of the unusually large growth stresses caused the wood to split apart as shown. The outward bending of the fragments confirms that the original surface residual stresses were tensile.

Figure 1.13: Growth stresses in a eucalyptus tree caused the trunk to split open after felling (photo courtesy of Wood Machining Institute).

Less dramatic, but also with very damaging and practically important effects, are the dimensional distortions that can occur in materials containing residual stresses during subsequent machining. Removal of stressed material locally redistributes the stresses in the remaining material so as to maintain internal force equilibrium. This stress change creates a corresponding strain change, which in turn causes an overall geometry change. Even small geometry changes can be very damaging when seeking to produce high precision components. In extreme cases—such as in Figure 1.14—where substantial and widespread residual stresses were present in the original billet and where a large percentage of material is removed, the geometric distortion can become very plainly visible.

1.6   RESIDUAL STRESS MEASUREMENTS

The dimension changes that occur when stressed material is removed from a test object provides a very effective means of measuring residual stresses. In general, the response to the removal of stressed material is linear elastic and so there is a unique relationship between the measured deformations and the residual stresses that originally existed in the removed material. It is then possible to evaluate those residual stresses by analysis of the observed deformations, measured either as strains or displacements. This is the basis of the so-called "destructive" measurement methods, also called "relaxation" methods because the material removal causes the stresses in the

Figure 1.14: C-17 cargo ramp warped by the release of residual stresses within the material machined away during manufacture (photo courtesy of D. Bowden, Boeing Company).

remaining material to relax partially. Chapter 2 describes many of these methods in more detail. The Hole-Drilling Method, the subject of this book, is the most well-known and commonly used relaxation/destructive method for measuring residual stresses.

1.7 FURTHER READING

- Ashby MF, Jones DRH (1980). *Engineering Materials—An Introduction to their Properties and Applications*, Pergamon Press, Oxford, UK.

- Withers PJ, Bhadeshia HKDH (2001). Residual Stress. Part 2—Nature and Origins, *Materials Science and Technology*, 17(4):366–375.

- Withers PJ (2007). Residual Stress and its Role in Failure, *Reports on Progress in Physics*, 70(12):2211–2264.

- Hosford WF (2009). Mechanical Behavior of Materials, 2nd ed., Cambridge University Press, Cambridge, UK, Chapter 18 in *Residual Stresses*.

- Niku-Lari A (Ed.) (1987). *Advances in Surface Treatments, 4: Residual Stresses*, Pergamon Press, Oxford.

- Noyan IC, Cohen JB (1982). The Nature of Residual Stress and its Measurement, Chapter 1 in *Residual Stress and Stress Relaxation*, Kula, E. (Ed.), Springer, New York.

- Noyan IC, Cohen JB (1991). Residual Stress in Materials, *American Scientist*, 79(2):142–153.

- Korsunsky AM (2017). *A Teaching Essay on Residual Stresses and Eigenstrains*, Butterworth-Heinemann, Oxford, UK.

- Totten G, Howes M, Inoue T (Eds.) (2002). *Handbook of Residual Stress and Deformation of Steel*, ASM International, Materials Park, OH.

- Hauk V (Ed.) (1997). *Structural and Residual Stress Analysis by Non-Destructive Methods*. Elsevier, Dordrecht, The Netherlands.

- Scafidi M, Pitarresi G, Toscano A, Petrucci, G, Alessi S, Ajovalasit A (2015). Review of Photoelastic Image Analysis Applied to Structural Birefringent Materials: Glass and Polymers. *Optical Engineering*, 54(8), 081206.

CHAPTER 2

Relaxation Type Residual Stress Measurement Methods

"According to Lord Kelvin's principle, the best evidence of residual stresses may be obtained by their measurement."

William Osgood (1955) in *Residual Stresses in Metals and Metal Construction.*

2.1   INTRODUCTION

Methods for measuring stresses, residual or applied, generally determine stress indirectly. Typically, they measure some proxy for stress, for example, strain, from which stress is determined using a "calibration." In simple cases, the needed step could just be a multiplication by Young's Modulus to relate measured strain to stress. However, in most cases a combination of detailed material property data and sophisticated computational procedures is required.

The "locked-in" character of residual stresses adds a further measurement challenge, independent of the measurement technique or calibration type used. For applied stresses, the typical procedure is to make comparative measurements on the structure before and after application of an external load, and then to evaluate stresses based on the difference between the measurements. The "no-load" state provides the zero datum. However, because of their locked-in nature, residual stresses cannot be analogously varied. In addition, identification of the zero-stress datum can become very challenging, often the greatest challenge. The various available residual stress measurement methods approach zero datum identification in different ways, and in all cases, great procedural care must be exercised to achieve success. A consequence of this issue is that residual stress measurements do not typically reach the accuracy or reliability that is possible when working with applied stresses. However, the various residual stress measurement methods are now quite mature and the accuracy gap is often not very large.

Residual stress measurement methods are typically divided into three general types:

- **Relaxation ("destructive") methods.** These are the subject of this chapter. They are the most generally applicable methods for a wide range of materials, both metallic and nonmetallic. They involve measuring the deformations caused by the cutting of some part of the stressed specimen. The resulting specimen damage can sometimes be quite extensive, but often is minimal and unharmful.

- **Diffractive methods.** These include the X-ray, Synchrotron and Neutron Diffraction methods. They involve measuring the diffraction patterns of the given radiation type from a stressed crystalline material. These measurements indicate the crystal lattice spacing, from which the local residual stresses can be evaluated. These methods have the advantage of being non-destructive, but are limited to crystalline materials. The X-ray technique can measure only very near-surface stresses. Synchrotron and Neutron Diffraction measurements can penetrate deeper but can only be done at a very small number of major nuclear facilities.

- **"Other" methods.** These include magnetic, ultrasonic, thermo-elastic and photoelastic methods. These are more specialized measurement methods suited to particular materials, based on specific properties of those materials and requiring detailed material-specific calibrations.

All the various methods have particular strengths that can make them well suited to certain measurement tasks, and also particular weaknesses that can make them unsuited to others. Thus, the method used for a given measurement task must be chosen with knowledge and care to achieve the best overall result. It is also important to understand and take into account possible associated limitations and concerns.

The Hole-Drilling Method, the subject of this book, is an example of a relaxation type method. It is a very versatile method and has a wide range of practical applicability. This chapter gives a general overview of the commonly used relaxation type residual stress measurement methods so as to place the Hole-Drilling Method in context. It describes the physical features and concepts that are important to understand when considering relaxation type methods in general and the Hole-Drilling Method in particular. In addition, while the Hole-Drilling Method is a good general-purpose choice, there are many cases where the use of an alternative method is appropriate. The larger viewpoint presented in this chapter is intended to assist with that choice.

2.2   RELAXATION METHOD CONCEPT

The relaxation methods for measuring residual stresses conceptually parallel the typical procedure used for measuring applied stresses. The method used for applied stresses is to measure the deformation change that occurs in the given specimen from either addition or removal of the applied load and then to evaluate the stresses based on the deformation measurements. The no-load case provides the zero-stress datum. In the case of residual stresses there is no external load to manipulate, so instead the "internal load" is manipulated. This is done by cutting or removing material containing residual stresses. The material cutting creates a stress-free surface, thus forming a zero stress datum locally. The residual stresses in the remaining material then redistribute themselves so as to re-establish internal force equilibrium. The deformations associated with this stress redistribution provide the data from which to evaluate the originally existing residual stresses. Since the material cutting locally reduces the original residual stresses, the various procedures used are called "relaxation" methods. The material cutting also damages the material to some extent, possibly substantially, so the name "destructive" methods is also often used. Figures 1.6, 1.13 and 1.14 in Chapter 1 illustrate the material deformations that arise when cuts occur in material containing residual stress. In these particular examples, the deformations are very large and dramatic, much larger than would typically occur in practical residual stress measurements.

2.3 EXCISION METHOD

The Excision Method provides the simplest conceptual example of a relaxation type method for evaluating residual stress. The procedure involves measuring the deformation of a small material fragment, typically using strain gauges, as it is cut from the bulk of the specimen material. Figure 2.1 illustrates the process. For these measurements it is assumed that:

1. The residual stresses are uniform over the size of the removed material. Conversely, the removed material size is small compared with the characteristic length of the residual stresses that are present (see Chapter 1, Section 1.4).

2. The process of cutting out the removed material does not introduce any new residual stresses. This issue often poses a significant experimental challenge.

3. The specimen material properties are reliably known. Commonly, Young's modulus and Poisson's ratio are needed to relate measured strains to stresses. For non-isotropic materials, directional elastic properties are needed, as well as the orientation of the principal material axes.

4. The quantitative form of the residual stress relief is reliably known and can be used to compute the original residual stresses from the measured deformations. In the case of excision, the stress relief is total, but for general relaxation type measurements the residual stresses are only partially relieved.

These four issues are common to all relaxation type measurements. Because of the full relief of the original residual stresses when doing excision, the stress evaluation described in item 4 above can be achieved by simple application of Hooke's Law. In addition, because the strain change corresponds to stress unloading, the calculated stresses and the measured strains have opposite signs. This feature is also shared by all relaxation type measurements.

Residual stress calculations for general relaxation type measurements are significantly more complex than for the Excision Method because:

1. The residual stresses are "relaxed" by the material cutting but are not entirely relieved. Thus, a detailed evaluation needs to be made to determine the specific fraction of the stress relief so as to provide proper scaling of the residual stress results.

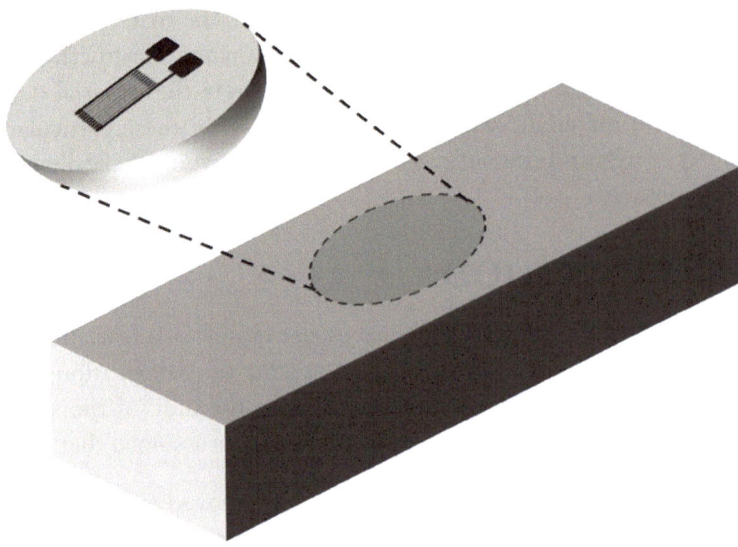

Figure 2.1: Excision Method for measuring residual stresses.

2. The stress release at a material cut occurs in one region of the specimen while the deformation measurements are made in an adjacent region. Thus, the residual stress calculation has to account for this spatial separation.

The Excision Method is exceptional because it avoids both issues by its complete relief of residual stresses in the same material as used for the deformation measurements. However, the large majority of relaxation type measurements involve one or both issues and therefore require more complex residual stress evaluation procedures. Over the years, substantial analytical and computational advances have been made, and have substantially mitigated the challenges created by the above two issues. Most methods are now quite well developed and can give reliable results when appropriately used.

2.4   TWO-GROOVE METHOD

This method, also called the "Two-Groove" or "Dual-Trench" Method, is a practical adaptation of the Excision Method. While conceptually simple, the Excision Method is challenging to implement in practice because it is difficult to cut a small test fragment without either damaging the fragment or introducing significant cutting stresses. The presence of a strain gauge increases the challenge. The Two-Groove Method is a practical expedient that avoids the need for complete removal of the test material from the larger specimen. In practice, only a partial removal is sufficient to release almost all the local residual stresses. The typical procedure is to attach a strain gauge on the specimen surface at the location of interest and to cut two deep slots on either side to release the residual stresses locally. The slots are made long and deep enough that the residual stresses originally existing at the strain gauge location are almost entirely relieved. A slot length > 4D and slot depth d > D is sufficient. In this way, the material attached to the strain gauge is conceptually excised, even though not physically. Figure 2.2 illustrates the process.

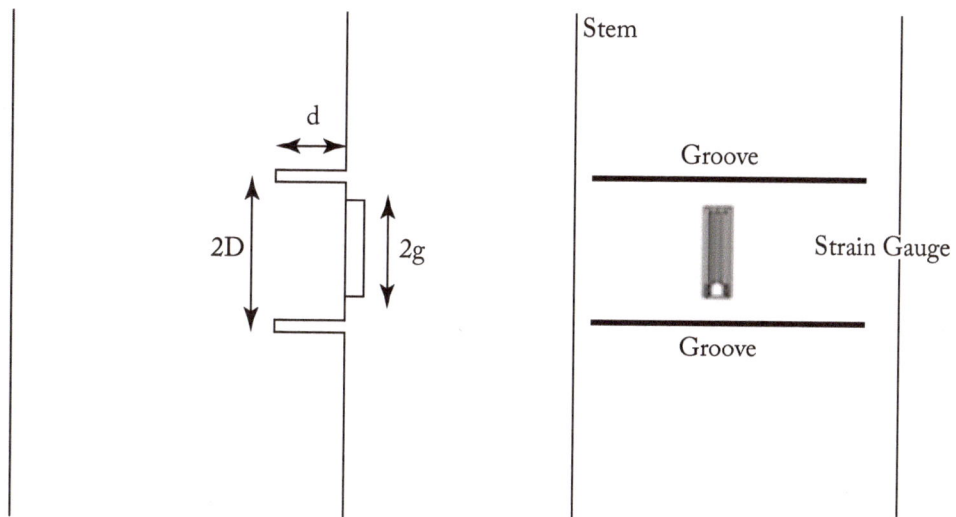

Figure 2.2: Schematic of Two-Groove Method (from Jullien and Gril (2007)).

The Two-Groove Method demonstrates fundamental characteristics that are typical of relaxation methods in general:

1. The released residual stresses are those that originally acted on the plane of the cut surface. These are the specific residual stresses that are evaluated by the measurement method. In the case of the Two-Groove Method, the main interest focuses on the normal in-plane stress acting perpendicular to the slot direction. Thus, for effective use of the method, the direction of the major in-plane residual stress must be known in advance so that the strain gauge and slots can be aligned accordingly.

2. The shear stresses that originally acted on the plane of the cut surface are also released. In the case of the Two-Groove Method, a perpendicular strain gauge such as shown in Figure 2.2 would not detect the associated shear strains, so the desired measurement of the normal stresses is not affected. If a shear stress evaluation is desired, a strain gauge aligned at 45° is required.

2.5   SPLITTING METHOD

The Splitting Method illustrated in Figure 2.3a has some geometric similarities to the Two-Groove Method, but typically uses only a single slot. The measurement concept derives from the deformations seen in material that has cracked due to excessive residual stresses, for example, as shown in Figures 1.6, 1.13 and 1.14 in Chapter 1. A deep cut is sawn into a specimen such as in Figure 2.3a and the opening or closing of the adjacent material indicates the sign and the average size of the residual stresses present. An opening motion indicates perpendicular residual stresses that are tensile on the outer surfaces and compressive in the interior, while closing indicates the opposite. This method is convenient as a quick comparative test for quality control during material production.

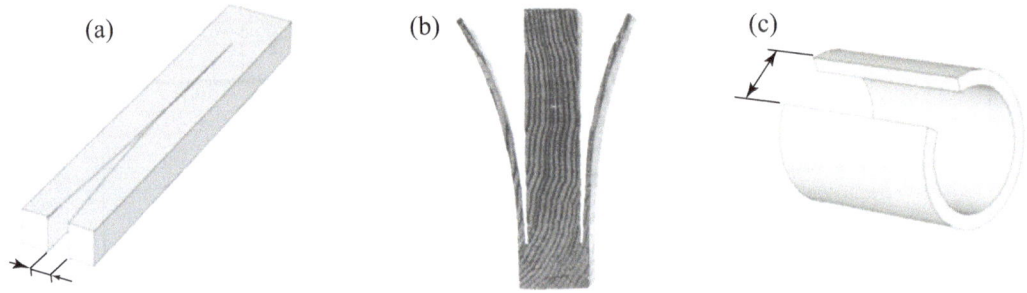

Figure 2.3: The splitting method, (a) for longitudinal stresses in rods, (b) for axial stresses in wood (from Fuller (1995)), and (c) for circumferential stresses in tubes.

A similar material response can occur when sawing wood. As seen in Figure 1.13 in Chapter 1, wood can contain significant residual stresses. In the case shown, these are "growth stresses," which are formed in the living tree to enhance its overall bending strength. Further substantial residual stresses can be created during the wood drying process, especially when accelerated by the use of drying kilns. Residual stress induced deformations can be problematic in sawmills when cutting the dried wood. If the material motion happens to be inward, the wood will close up on the sawblade during cutting, possibly quite tightly. Should the wood workpiece not be secured sufficiently, the saw will grab the wood and throw it out from the machine at high speed, creating a serious safety hazard. Thus, it is common to test for the presence of residual stresses in kiln-dried lumber. This is done using the "prong" test illustrated in Figure 2.3b.

An interesting variation of the prong test involves using a specimen similar to that shown in Figure 2.3a and to measure the lateral force required at the open end to maintain the original spacing between the two cut sides. This use of a force measurement instead of a displacement measurement is of interest because it gives a surface stress result that does not require explicit knowledge of Young's modulus. This feature is attractive when working with a natural material like wood whose physical properties are highly variable and typically are not accurately known without further testing.

Figure 2.3c shows another geometrical variant of the Splitting Method, commonly used to assess the circumferential residual stresses in thin-walled heat exchanger tubes. Diameter increase caused by the opening of the cut indicates tensile circumferential stresses around the exterior surface of the tube, balanced by compressive stresses around the interior. Diameter decrease indicates the opposite. The experimental method is well-established and is specified by ASTM Standard Test Procedure E1928. The approximate sizes of the surface circumferential stresses can be estimated using:

$$\sigma_\theta = \pm \frac{Et}{1 - \nu^2} \frac{D - D_0}{DD_0},$$

(2.1)

where σ_θ are the circumferential bending stresses at the outer $(+)$ and inner $(-)$ surfaces, D_0 and D, respectively, are the diameters before and after making the cut, t is the wall thickness, E is Young's modulus and ν is Poisson's ratio. The calculated results are approximate because Equation (2.1) is based on an assumption of linearly varying bending stresses through the tube wall thickness. In practice, the residual stresses do not vary exactly linearly.

2.6   SLITTING (CRACK COMPLIANCE) METHOD

From a conceptual point of view, the Slitting Method, schematically illustrated in Figure 2.4, is a further variant of the Two-Groove Method. In this case, relieved strain measurements are sequentially made as slot cutting proceeds in a series of small incremental steps. The set of strain measurements provides sufficient data for the evaluation of the stress profile within the slot depth. This process differs from the typical Two-Groove measurement, where just a single before-and-after strain measurement is made while cutting the slots in just one step directly to the steady-state depth. This latter measurement gives only the weighted average stress within the cut depth, with no within-depth profile information.

The purpose of using two deep slots in the Two-Groove Method is to create full strain relief in the enclosed material, thereby providing a simple residual stress evaluation. Since a stress profile measurement using intermediate slot depths involves measurement and analysis of partial strain reliefs, full strain relief never occurs except perhaps at the very end. Thus, the second slot does not provide any computational advantage and so is omitted to simplify the required experimental procedure. In addition, as shown in Figure 2.4, the strain gauge position is not limited to being on the specimen top surface. Other locations are also useful, notably on the opposite surface of the material specimen. As a general rule-of-thumb, strain measurements are most sensitive to nearby stresses. Thus, the top and bottom surface strain gauges shown in Figure 2.4 are useful for determining stresses near their respective locations, thereby achieving better spatial coverage.

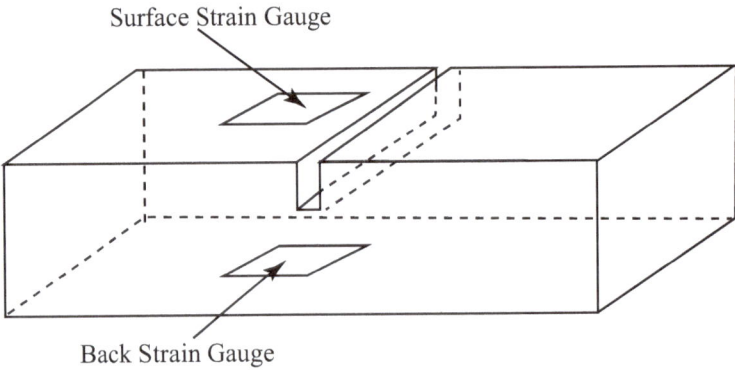

Figure 2.4: Slitting Method for measuring residual stresses.

Despite its geometrical similarity to the Two-Groove Method, the historical roots of the Slitting Method have a separate origin based on an earlier analogy with the strain field around a crack. The crack analogy provided the needed computational procedure to enable evaluation of the within-depth stress profile from the set of incrementally measured strains. Based on this approach, the original procedure name was the Crack Compliance Method. However, the stress computational procedure is now typically based on numerical calibration data from finite element calculations, so the crack analogy is no longer needed. Consequently, the name "Slitting Method" or "Incremental Slitting Method" has come into more common use.

The typical data from the Slitting Method are the set of strain gauge measurements acquired as the depth of the slit is incrementally increased in small steps. The surface strain $\varepsilon(h)$ measured when the slit reaches a depth h is the sum of the relaxation effects of all the perpendicular normal stresses at the various depths within the slit depth:

$$\varepsilon(h) = \int_0^h A(H, h)\, \sigma(H)\, dH, \tag{2.2}$$

where $\sigma(H)$ is the normal stress existing at depth H from the surface and $A(H, h)$ is the strain response from a unit stress existing at depth H within a slit of current depth h. The compliance function $A(H, h)$ is typically determined using finite element calculations. The integral on the right side appears as a consequence of all the residual stresses over the cut surface combining to produce the measured strains.

Equation (2.2) is mathematically classified as a Volterra equation of the first kind. It is an "inverse" equation because the quantities to be determined, the stresses, appear on right side within the integrand, while the known data, the strains, appear on the left. This is the reverse of the usual "forward" format. The inverse character of Equation (2.2) substantially complicates the required solution procedure and causes the stress results to be sensitive to the presence of small strain measurement errors. Sophisticated mathematical procedures are required to get reliable stress solutions.

The Slitting Method and Equation (2.2) provide a conceptual prototype for the majority of the relaxation type of residual stress measurement methods. The common features of these methods are:

1. Deformations, typically strains, are measured at one or more points on the specimen while the cut surface is deepened into the interior in a series of small steps. There is a spatial separation between the measurement location and the associated stresses within the cut depth. Thus, the relationship between the residual stresses and the measured strains is indirect and requires either analytical or numerical "calibration." This feature is represented in Equation (2.2) by the compliance function $A(H, h)$ also called the "kernel" function.

2. The out-of-plane residual stresses that originally existed along the cut surface within the cut depth all contribute to the measured strains. Because many stresses contribute to each measured strain, there is not a one-to-one relationship between individual strains and stresses. This feature is represented in Equation (2.2) by the presence of the integral on the right side.

3. The measured strain is mostly controlled by the residual stresses nearest to the measurement location. In mathematical terms, the measured strain response is a weighted average from the stresses that originally existed over the full cut depth, where the weighting is heavily biased toward the measurement location.

4. When making measurements at the top surface, the measured strain gradually increases (and then sometimes slightly decreases) as the cut is extended, reaching a steady-state level when the cut extends far from the strain measurement location. If the residual stresses are assumed to have uniform variation within the cut depth, measurement of only the steady-state strain provides a convenient way to determine the uniform residual stress. This is the procedure typically used with the Two-Groove Method, where the strain measurement is made after cutting the slots in one step to a depth at least reaching the steady-state strain value, which approximately corresponds to the slot separation distance. For the Tube Splitting Method, the equivalent uniform bending stresses are estimated from the tube diameter change before and after full splitting of the tube.

5. The general format of Equation (2.2) applies to a wide range of relaxation type methods for measuring residual stresses. The various methods differ substantially in their specimen and cut geometries, but mathematically, the associated residual stress calculations fit the format of Equation (2.2), each method with its own specific compliance function $A(H, h)$. Some methods may involve measurements of displacements rather than strains, sometimes also at more than one point. The details of Equation (2.2) then vary accordingly, but the overall format remains the same. Figure 2.5 shows example contour plots of cumulative

compliance functions for several common residual stress measurement methods. Plotted vertically is the cut depth h, and horizontally the stress depth H. Because only the stresses within the cut depth can contribute to the strain response, H is always less or equal to h. This is the reason for the triangular shape of the plots in Figure 2.5 and also for the presence of h as the upper limit of the integral in Equation (2.2). It can be seen that the various cumulative compliance functions differ in detail, but have similar overall characteristics.

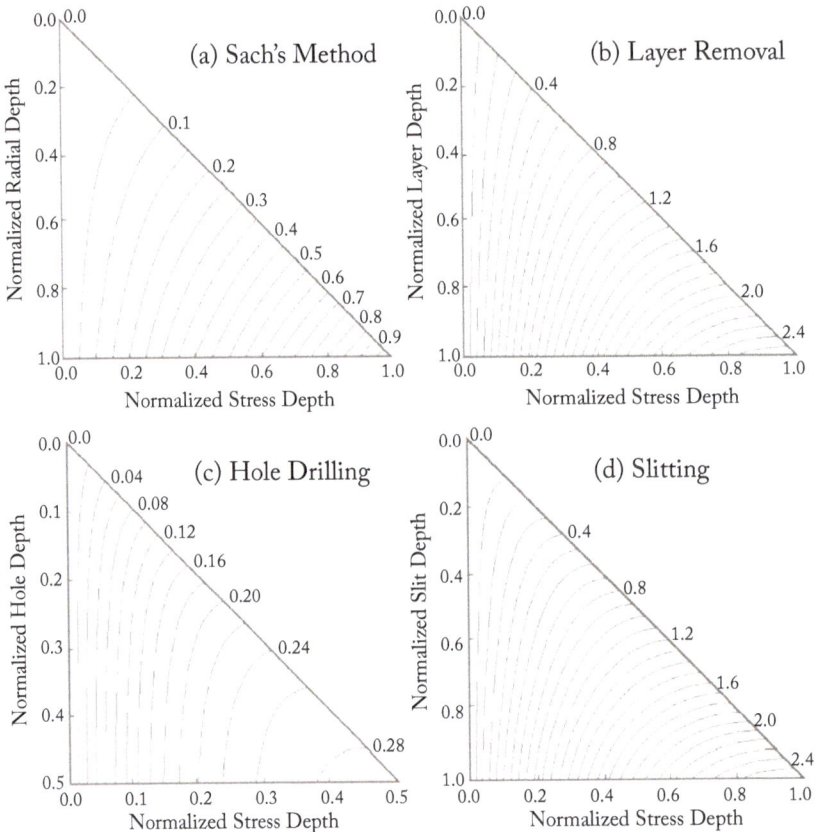

Figure 2.5: Cumulative compliance functions for various residual stress measurement methods. (a) Sachs' Method, (b) Layer-Removal Method, (c) Hole-Drilling Method, and (d) Slitting Method (from Schajer and Prime (2006)).

6. A characteristic of an integral equation such as Equation (2.2) is that very different combinations of stresses can sum together to produce almost similar corresponding strains. This characteristic means that small errors in the measured strains can substantially shift the details of associated stress solution. The mathematical effect is like a "noise amplifier," where small noise in the input strain data creates proportionally much larger noise in the

output stress results. While the effect is seen in the mathematics, the behavior has physical origins. It occurs because of the indirect relationship between strain and stress that occurs due to their spatial separation. This is a St. Venant type effect, which gets worse as the separation between the strain and stress locations increase. Thus, for the Slitting Method, for example, the ability to identify interior stresses from a top-surface strain gauge measurement rapidly diminishes with increasing stress distance from the strain gauge. Over the years some sophisticated mathematical methods have been developed to stabilize the stress solutions in the presence of strain measurement noise. These are effective, but they cannot circumvent the underlying physics and they cannot give solutions for interior stresses at remote cut depths beyond the point where those stresses no longer make any significant contribution to the measured strains. In addition, the mathematical methods can only reduce sensitivity to measurement noise, not eliminate it. Thus, a scrupulous experimental procedure remains essential to minimize the presence of noise in the original measured data.

2.7   RING-CORE METHOD

The Ring-Core Method is in concept a 2-D extension of the Two-Groove and Slitting Methods. The latter two methods are both single-direction methods that measure the normal stress perpendicular to the slots or slit. This characteristic requires that the direction of the most significant normal stress be known in advance so that the cut direction can be chosen appropriately. In-plane shear stress could also be evaluated by using an additional 45° strain measurement, although this is rarely done in practice. For the general 2-D case the Ring-Core Method, schematically illustrated in Figure 2.6, is used. Here, the slots of the Two-Groove Method are adapted into a single circular slot that isolates the enclosed island of material in much the same way as the Two-Groove Method, but from all in-plane directions. Thus, it is not necessary to know or guess in advance the desired measurement direction. The single strain gauge of the Two-Groove Method is replaced by a three-element rosette to provide sufficient data to evaluate the two in-plane normal stresses and the in-plane shear stress. The Ring-Core Method is quite widely used and several specialized strain gauge rosettes and annular milling devices are available commercially.

The Ring-Core Method can be used in two different ways. In the first way, the method is equivalent to a 2-D version of the Two-Groove Method. Strain readings are taken before and after ring cutting in one step to a depth sufficient to relieve the surface residual stresses completely. The three in-plane stress components can then be evaluated from the measured strain changes by direct application of Hooke's Law. This type of measurement requires ring cutting to a depth approximately equal to the ring diameter, commonly ~20 mm but could be greatly smaller or larger. This type of measurement is suited to the case where the residual stresses are known to be generally uniform with depth from the measured surface. The evaluated stresses are average values, weighted in favor of the near-surface stresses.

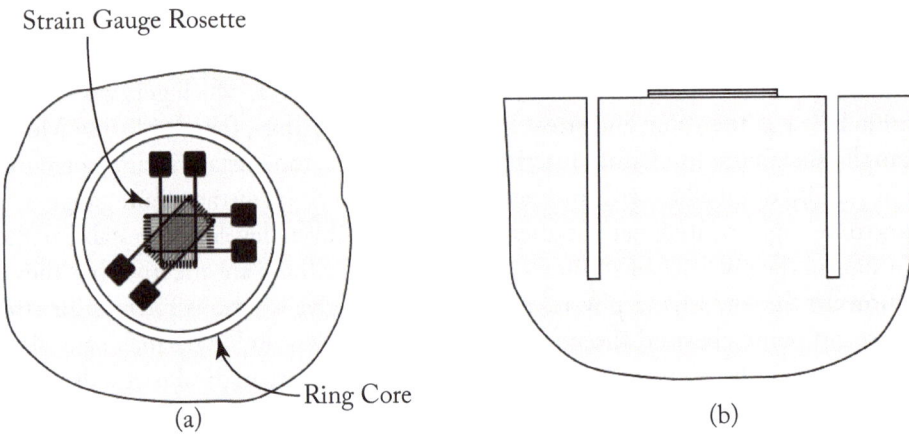

Figure 2.6: Ring-Core Method. (a) Plan view and (b) cross-section view.

The second way of using the Ring-Core Method is equivalent to a 2-D version of the Slitting Method. In this case, sets of relieved strain readings are sequentially taken as the ring depth is progressively increased in a series of small steps. These strain readings provide the input data for three integral equations similar to Equation (2.2), each specialized to a component of the three in-plane residual stresses. The mathematical procedure is the same as for the Hole-Drilling Method and is described in more detail in Chapter 5. In this way, the within-depth profile of the in-plane stresses can be determined to a depth approximately equal to the ring radius. Residual stresses at greater depths make rapidly decreasing contributions to the relieved strains at the surface and cannot be resolved reliably.

Figure 2.6 illustrates the classical approach for doing Ring-Core residual stress measurements using strain gauges. In recent years, full-field optical techniques have been introduced for measuring the surface deformations. These techniques enable measurements to be made over a wide range of dimension scales, including microscopic. Of interest to note is that while the strains that occur in Ring-Core measurements are greater than for the Hole-Drilling measurements described below, the surface displacements that occur in both methods have similar sizes for a given annulus/hole diameter. This occurs because the region within the annulus has large strain but small radius, while the area outside the annulus/hole has smaller local strains but larger radius. The two effects approximately balance each other, thereby giving similar displacements in the two regions. Chapter 7 describes some commonly used optical measurement methods.

2.8 HOLE-DRILLING METHOD

The Hole-Drilling Method, illustrated in Figure 2.7, is similar in concept to the Ring-Core Method, but with reversed geometry. While the Ring-Core Method involves making deformation measurements in the central area while the surrounding material is cut away, the Hole-Drilling Method involves making deformation measurements in the surrounding area while the central material is cut away. This similarity means that the two methods share many similar characteristics and can use analogous calculation procedures to evaluate the residual stresses from the measured deformation data.

In the classical Hole-Drilling Method, a specially designed strain gauge rosette is attached to the specimen at the desired location. A hole is then incrementally drilled in a series of small steps at the geometric center of the rosette, with strain readings taken after completion of each depth step. Analysis of the relieved strain measurements enables the through-thickness profiles of the three in-plane stress components. These stress profiles can be determined to a depth approximately equal to the hole radius. Three sizes of strain gauge rosettes are commonly available to accommodate holes of nominal diameter 1/8", 1/16", and 1/32" (3 mm, 1.5 mm, 0.75 mm). Choice of rosette size used depends on the size of the specimen and the desired depth for the stress profile evaluation.

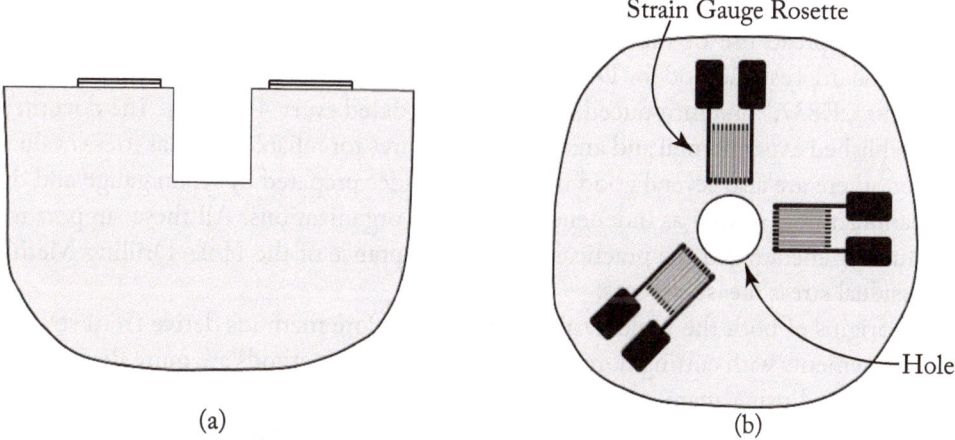

(a)   (b)

Figure 2.7: Hole-Drilling Method. (a) Cross-section view and (b) plan view.

The choice between the Hole-Drilling and Ring-Core methods mostly depends on procedural convenience and user preference. Of the two, the Hole-Drilling Method is by far the more widely used, possibly also the most widely used among all residual stress measurement methods. The reasons for this are the straightforward experimental procedure, the wide range of applicable material types and specimen geometries, the relatively modest cost of the experimental equipment and the reliability of the residual stress results. In addition, the drilled hole is often sufficiently small compared with the specimen size that its presence does not significantly impair specimen functionality. For this reason, the Hole-Drilling Method is sometimes called "semi-destructive." By comparison, the diameter of the annular groove used for the Ring-Core

Method is typically ∼20 mm, is about an order of magnitude larger than for the Hole-Drilling Method, thereby causing correspondingly larger specimen damage.

The experimental simplicity of the Hole-Drilling Method is enhanced by the external arrangement of the strain gauges around the hole. This means that the hole can be cut while keeping clear of the strain gauges and their associated wiring. In contrast, the strain gauge wires in the Ring-Core Method cross the cut path for the annular groove, thus demanding significant operator care to avoid damaging the wires.

One area when the Ring-Core Method may be chosen over the Hole-Drilling Method is for the measurement of residual stresses that are close to the material yield stress. The presence of the hole creates a stress concentration that can cause local material yielding during hole-drilling. This stress concentration exists around the outside of the hole and so affects only Hole-Drilling measurements. By contrast, the interior of the annular groove made for Ring-Core measurements does not have stress concentrations and so is uninfluenced. Using conventional analytical procedures, Hole-Drilling measurements can measure residual stresses up to 80% of yield stress. Chapter 5 describes specialized analytical procedures that can allow measurements beyond 90% of yield stress.

The widespread use of the Hole-Drilling Method has led to the development of the "ASTM Standard Test Method for Determining Residual Stresses by the Hole-Drilling Strain-Gage Method, E837," first introduced in 1985 and updated every 4–6 years. The document describes established experimental and analytical procedures for reliable residual stress evaluations. In addition, there are also several good instruction guides prepared by strain gauge and drilling device manufacturers as well as independent research organizations. All these support materials have further encouraged the practical use and acceptance of the Hole-Drilling Method for general residual stress measurements.

The origins of both the Hole-Drilling and Ring-Core methods derive from strain gauge based measurements with cutting done on a mm scale. Both methods are quite flexible and they are also conducted using many other measurement techniques, notably optical, and on scales ranging from microscopic to major structural size. These features are discussed in more detail in subsequent chapters.

2.9   DEEP-HOLE METHOD

A general characteristic of relaxation type residual stress measurement methods is that they are mostly sensitive to the stresses near the measurement location. Their sensitivity to remote stresses rapidly diminish with distance. This effect is a consequence of the action of St. Venant's Principle and must be taken into account when choosing or designing a residual stress measurement method.

The Hole-Drilling and Ring-Core methods both use deformation measurements made on the specimen surface. Consequently, they are mostly sensitive to near-surface stresses, with rapidly diminishing sensitivity to deep interior stresses. If evaluation of deep interior stresses is desired, it is necessary to move the deformation measurement location to the region of interest. The Deep-Hole Method, illustrated in Figure 2.8, uses this strategy. It involves drilling an initial reference hole, placing a sensor in the hole at the desired interior region, and making deformation measurements, typically diameter change, as the surrounding material is separated from the bulk material by annular over-coring.

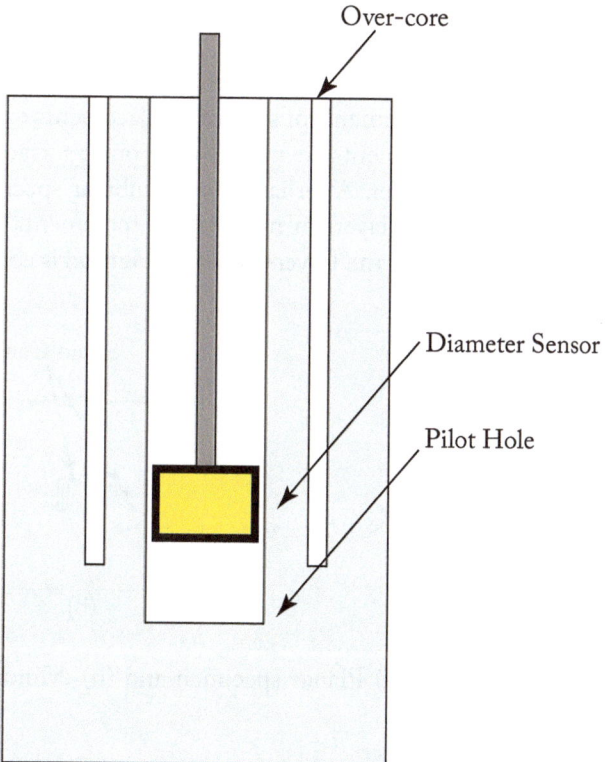

Figure 2.8: Cross-section view of the Deep-Hole Method.

Measurement of diameter change in at least three directions allows evaluation of the local perpendicular normal and shear stresses. An interesting feature of the Deep-Hole Method is that the cutting of the initial reference hole creates a stress concentration in the surrounding material. The subsequent over-coring relieves these locally increased stresses, not just the originally existing residual stresses. Thus, the Deep-Hole Method has a greater stress sensitivity than would be expected from direct relaxation of the original residual stresses. However, the stress concentration can also cause localized material yielding when large residual stresses, close to the material yield stress, are present. This effect must be accounted for to get reliable results.

The Deep-Hole Method was originally introduced for geological use as a means of measuring stresses within large rock masses. For this application, the reference hole can reach several hundred meters deep. On a smaller but still substantial scale, the Deep-Hole Method is also used for measuring residual stresses in large metal castings at depths of several decimeters.

2.10 LAYER-REMOVAL METHOD

The Layer-Removal Method, schematically illustrated in Figure 2.9, is a long-established method for measuring residual stresses in planar and cylindrical specimens. For planar specimens, it involves measuring the deformation on one surface as layers of material are incrementally removed from the opposite surface. The deformation measurements are typically done using strain gauges or by optical measurements of surface displacements or curvature. For cylindrical specimens, deformation measurements are made on the outer surface as layers of material are incrementally drilled from the center. Alternatively, for tubular specimens, measurements can be made on the interior surface as layers of material are incrementally removed from the outer surface. For cylindrical specimens, the Layer-Removal Method is commonly called Sachs' Method.

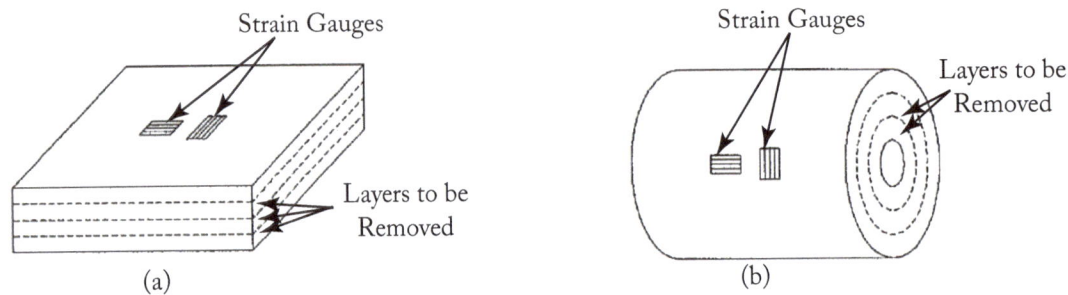

Figure 2.9: Layer-Removal Method. (a) Planar specimen and (b) cylindrical specimen (from Schajer (2001)).

The Layer-Removal Method well exemplifies the action of the governing integral Equation (2.2). For example, for a tubular specimen, the relationship between the strains measured on the interior surface as layers of material are incrementally removed from the outer surface is:

$$\varepsilon^*(r) = \frac{1 - \nu^2}{E} \int_a^r \frac{-2r}{r^2 - a^2}\sigma(R)\,dR \tag{2.3}$$

where

$$\varepsilon^*(r) = \varepsilon_\theta(r) + \nu\varepsilon_a(r) \tag{2.4}$$

and where $\varepsilon_\theta(r)$ and $\varepsilon_a(r)$, respectively, are the circumferential and axial strains measured on the interior surface when the radius of the outer surface is r, a is the radius of the inner measurement surface, E is Young's modulus and ν is Poisson's ratio. The kernel function $-2r/(r^2 - a^2)$ corresponds to $A(r, R)$, where the original h and H in Equation (2.2) are here replaced by r and

R. The leading minus sign occurs because the measurement is made during stress unloading. The triangular contour plot in Figure 2.5a corresponds to this kernel function.

Stoney's Method is an interesting variation of the Layer Removal Method, typically used for evaluating the residual stresses in electroplated surfaces. It commonly happens that substantial in-plane residual stresses are developed during electroplating of one metal on to another, for example, to produce a corrosion-resistant surface. Nickel plating is a common example of such an application. However, the highly tensile residual stresses that are developed make nickel an undesirable choice for plating fatigue sensitive components.

Stoney's Method is a conceptual reversal of the Layer Removal Method insofar as it involves *addition* of a layer of stressed material rather removal. Mathematically, the only effect of this is to remove the minus sign from the residual stress calculation. In practice, Stoney type measurements are done by electroplating a thin layer onto one surface of a substrate plate and measuring the resulting plate curvature caused by the unsymmetrical loading. Alternatively, the deformation may be monitored using strain gauges fixed on the unplated side. The plating stress σ can then be determined from the measured surface strain ε using:

$$\sigma = \frac{E\,h\,\varepsilon}{3(1-v)\,t}, \tag{2.5}$$

where t is the plating thickness, h is the substrate plate thickness, E and v, respectively, are the Young's modulus and Poisson's ration of the substrate material. The factor $(1-v)$ occurs because the plating stresses exist equally in both in-plane axial directions. Use of Equation (2.5) requires that the plating is thin compared with the substrate plate, $t \ll h$, and that the substrate plate is in turn thin compared with its in-plane dimensions.

2.11 CONTOUR METHOD

The various measurement methods described so far have required that the residual stresses have a particular spatial distribution or to be uniform over significant distances. For example, for the Splitting Method, Figure 2.3a, the residual stresses are assumed to vary linearly across the cut ligaments. Similarly, for the Layer Removal method, Figure 2.9a the residual stresses need to have the same through-thickness profile over the entire surface area of the measured plate. These methods therefore give either average stress values or one-dimensional profiles. The Contour Method, illustrated in Figures 2.10a–c, is notable among the relaxation type measurement techniques in that it can provide a detailed two-dimensional map of the normal residual stresses acting on a plane within the specimen. The procedure involves cutting through the specimen cross-section using a wire Electro-Discharge Machining (EDM), and measuring the surface height maps of the cut surfaces using a coordinate measuring machine or a laser profilometer. The residual stresses shown in Figure 2.10a are released by the cut and cause the material surface to deform (pull inward for tensile stresses, bulge outward for compressive stresses), as shown in Figure 2.10b. The originally existing residual stresses normal to the cut can be evaluated from finite element calculations by determining the stresses required to return the deformed surface shape to a flat plane. The surface deformations are very small, so the EDM cutting and the surface height map measurement must be done to very high precision. In addition, to avoid any asymmetry effects, it is desirable to measure the surfaces on both sides of the cut and to use the average surface height map. It is also possible to make further cuts on perpendicular planes to get maps of the normal residual stresses on those planes. Figure 2.10d shows an example measurement of the axial residual stress profile within the cross-section of a railway rail.

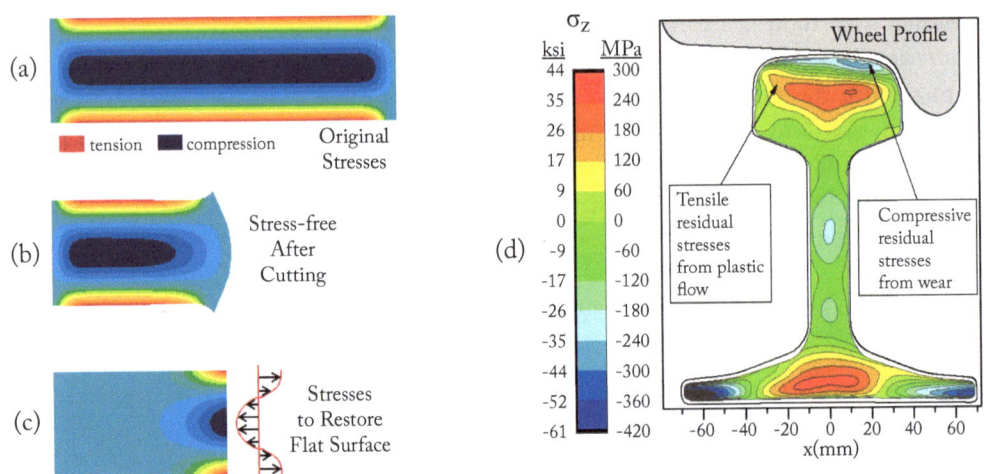

Figure 2.10: Contour Method. (a) Original stresses, (b) stress-free after cutting, (c) stresses to restore flat surface, and (d) measured stress profile of a railway rail (diagrams courtesy of Michael Prime, Los Alamos National Labs, USA).

2.12   SECTIONING METHOD

The previously described measurement methods each have a particular geometrical arrangement designed to achieve a specific type of measurement. The Sectioning Method is not so much a "method" as a generic approach that can be highly customized to fit specific individual specimens and measurement needs. The approach is to choose a measurement procedure that combines one or more of the previously described measurement methods and to use them sequentially to reach the desired measurement needs. The chosen measurement procedure must be selected with care to fit and perhaps also to exploit particular features of the specimen geometry. The Sectioning Method involves making deformation measurements at multiple steps within the sectioning process and at multiple locations. Any suitable deformation measurement technique can be used, although most reported Sectioning Method examples tend to be dominated by strain gauges.

Figure 2.11 schematically illustrates a typical Sectioning Method measurement. In this example, the sequence of cuts used resembles a combination of the Excision Method and the Layer Removal Method. A feature of the Sectioning Method is that the surfaces created by the cuts made in one step are available for use as measurement locations in subsequent steps. As can be seen in the example procedure, the Sectioning Method is highly destructive and typically the specimen is reduced to a large number of small pieces. However, the large amount of cutting allows for residual stress measurements to be made at multiple points of interest within the original specimen.

Residual stress computations for the Sectioning Method can often be challenging because each successive cut partially relieves the local residual stresses, leaving an altered residual stress state to be relieved further by subsequent cuts. Thus, the residual stress computation procedure has to keep track of this sequence so that the final results relate back to the originally existing residual stresses. This is not an easy task and has to be done with significant care. If done well, the Sectioning Method is a very flexible, if rather labor-intensive, method for measuring detailed residual stresses in a broad range of specimens.

2.13   IMPACT OF MODERN MEASUREMENT TECHNOLOGIES

Most of the residual stress measurement methods described in this chapter are long established with substantial histories. They can be implemented using various experimental techniques, chosen according to desired features. Over the years, as technology has advanced, the variety and functionalities of experimental techniques have greatly increased both in terms of procedural capability and precision. Such is the scale of these advances that the modern residual stress measurement procedures are essentially "new" methods when compared with the early versions. Modern computer-based computation methods have similarly revolutionized residual stress computation capabilities, allowing stress evaluations that were far beyond reach in earlier times. Figure 2.12 illustrates examples of both of these advances. The figure shows relaxation measurements made using various of the techniques described in this chapter, but done at the microscopic scale. The "cuts" shown in the images are made using a Focused Ion Beam within a Scanning Electron Microscope (FIB-SEM). The cut features are less than 0.5 μm wide, while the whole image width spans 15 μm. These are very small dimensions, the first equals the wavelength of visible (green) light, while the second is about one-fifth the diameter of a human hair. Measurements on this scale are useful for studying residual stresses within integrated circuits, MEMS devices and within individual crystal grains in metals.

The speckled background within the images in Figure 2.12 is a random dot pattern deliberately "decorated" onto the specimen surface by ion deposition. These speckles provide the basis for Digital Image Correlation (DIC) evaluation of the surface deformations caused by the various cuts into the stressed surfaces. The use of DIC in this way provides a very effective method for measuring surface displacements, even at this microscopic scale. Such DIC measurements

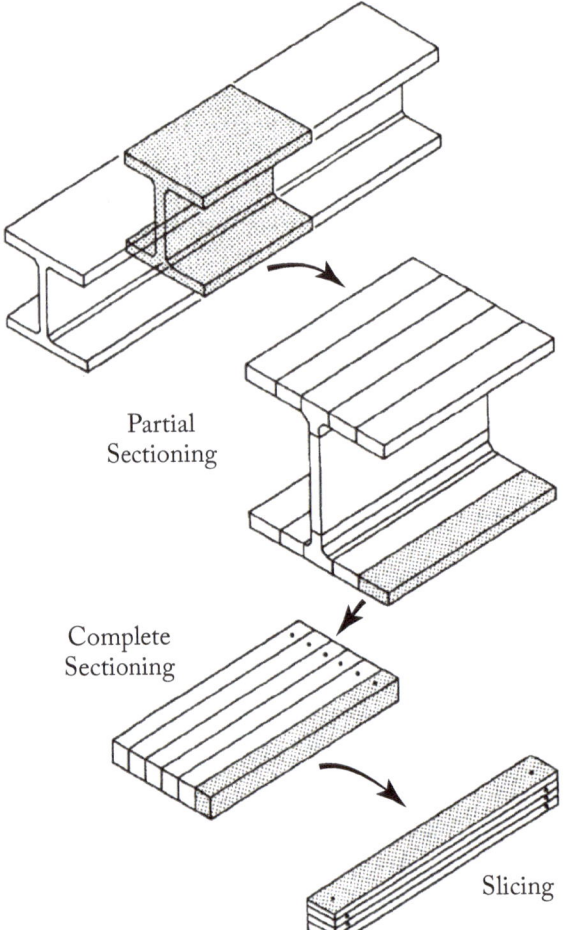

Figure 2.11: Sectioning Method example (from Tebedge et al. (1973)).

are an example of the way in which modern computation methods have expanded the scope of the traditional residual stress measurement techniques. Before the advent of laboratory computer capabilities, such calculations were totally impractical. Chapter 7 describes the use of modern optical measurement methods, including DIC, for residual stress measurements.

2.14 METHOD SELECTION

The residual stress measurement methods described in the previous sections span a wide range of potential specimen geometries and stress measurement features. Each method has particular strengths and limitations. Some measurement tasks can be accomplished using any of several

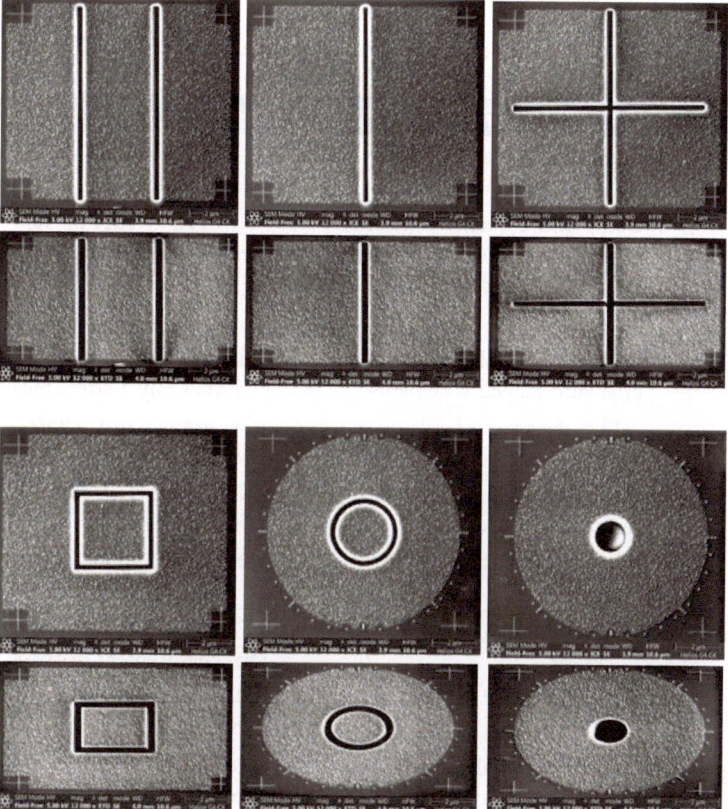

Figure 2.12: FIB-SEM micrographs of residual stress measurements done at a microscopic scale. The width of each image is ∼ 15 μm. Rows 1 and 3 are plan views and rows 2 and 4 are corresponding oblique views (images courtesy of Dr. B. Winiarski, Manchester University, UK).

methods, while others will be best achieved using a specific measurement method. Table 2.1 summarizes some general comments about the various stress relaxation methods discussed in this chapter. In addition, because particular measurement tasks may require further choices, mention is also made of the Diffractive and "Other" methods listed in Section 2.1. Detailed information about these methods can be found in the references listed in the "Further Reading" section.

Of the various techniques listed in Table 2.1, the Hole-Drilling Method stands out prominently as a "standard" general-purpose method for measuring residual stresses, perhaps the most commonly used residual stress measurement technique. The reasons for this popularity are its wide-ranging applicability, its ease and low cost of application and its reliable results. The following chapters describe the method in detail, with wide-ranging coverage of practical measure-

Table 2.1: Characteristics of various residual stress measurement methods

Method	Works Best For:	Limitations/Challenges:
Excision	Surface measurements where the stress direction is known	Difficult to cut material sample cleanly, without stress addition
Splitting	Comparative quality control of prismatic or tubular specimens	Gives an average or "representative" result only
Two-Groove	Surface measurements where the stress direction is known	Gives an average result only
Slitting	1-D perpendicular stress in prismatic shaped specimens	Stresses must be uniform across slit width. Only 1 stress measured
Ring-Core	2-D surface stresses, also near-surface stress profile	Creates much specimen damage, awkward strain gauge placement
Hole Drilling	2-D surface stresses, also near-surface stress profile	Near-yield stresses are overestimated
Deep Hole	Large components	Done only by specialists and compromised by plasticity
Layer Removal	Flat plates and cylinders of uniform thickness	Time consuming procedure, subject to measurement drift
Stoney	Thin layers on flexible substrate	Determining layer thickness accurately
Contouring	2-D perpendicular stress in prismatic shaped specimens	Requires very accurate cutting, not good for near-surface
Sectioning	Can be tailored to specific specimen geometry	Challenging calculations for multiple sectioning
X-ray Diffraction	Near surface measurements on crystalline materials	Variations in grain structure and surface texture
Synchrotron Diffraction	Deeper non-destructive measurements	Requires synchrotron radiation source in a major facility
Neutron Diffraction	Very deep non-destructive measurements	Requires neutron radiation source in a major facility
Magnetic (Barkhausen)	Rapid measurements in ferromagnetic materials	Requires extensive material-specific calibration
Ultrasonic	Low-cost comparative measurements	Requires material-specific calibration
Thermo-elastic	Low-cost comparative measurements	Results are not quantitative
Photo-elastic	Full-field measurements in transparent materials	Through-thickness average. Stress separation is challenging

ment and computational techniques. The Hole-Drilling Method is still very much a living and evolving method, with many significant advances made in recent years and more in progress. The later chapters in this book describe many of these developments.

2.15 FURTHER READING

Residual Stress Measurements

- Schajer GS (Ed.) (2013). *Practical Residual Stress Measurement Methods*, Wiley, West Sussex, UK.

- Totten G, Howes M, Inoue T (Eds.) (2002). *Handbook of Residual Stress and Deformation of Steel*, ASM International, Materials Park, OH.

- Kandil FA, Lord JD, Fry AT, Grant PV (2001). *A Review of Residual Stress Measurement Methods—A Guide to Technique Selection*. Report MATC(A)O4NPL, NPL, Teddington, Middlesex, UK.

- Withers PJ, Bhadeshia HKDH (2001). Residual Stress, Part 1: Measurement Techniques, *Materials Science and Technology*, 17(4):355–365.

- Schajer GS (2001). Residual Stresses: Measurement by Destructive Methods. In Buschow KHJ. et al. (Eds.), *Encyclopedia of Materials: Science and Technology, Section 5a*, Elsevier Science, Oxford.

- Osgood W (1955). *Residual Stresses in Metals and Metal Construction*. Reinhold Publishing Corp. New York.

- Schajer GS, Prime MB (2006). Use of Inverse Solutions for Residual Stress Measurements. *Journal of Engineering Materials and Technology*, 178(3):375–382.

Excision Method

- Scott P (1986). Comparison of the Blind-Hole Drilling and the BCD Chip-Removal Technique for Determining Residual Stresses. *Proc. SEM Spring Conference on Experimental Mechanics*, pp. 209–215, New Orleans, LA, June 8–13.

Two-Groove Method

- Schwaighofer J (1964). Determination of Residual Stresses on the Surface of Structural Parts, *Experimental Mechanics*, 4(2):54–56.

- Schwartz IF (1995). Multiple Trenching: A Simple Self-Calibrating Method of Residual-Stress Measurement, *Experimental Mechanics*, 35(3):201–204.

- Jullien D, Gril J (2008). Growth Strain Assessment at the Periphery of Small-Diameter Trees using the Two-Grooves Method: Influence of Operating Parameters Estimated by Numerical Simulations, *Wood Science and Technology*, 42(7):551–565.

Splitting Method

- ASTM (2013). Standard Practice for Estimating the Approximate Residual Circumferential Stress in Straight Thin-walled Tubing, E1928-13, ASTM International, West Conshohocken, PA.

- McMillen J (1958). Stresses in Wood During Drying. Report 1652, U.S. Department of Agriculture, Forest Service, Forest Products Laboratory. Madison, WI.

- Diawanich P, Matan N, Kyokong B (2010). Evolution of Internal Stress During Drying, Cooling and Conditioning of Rubberwood Lumber, *European Journal of Wood and Wood Products*, 68(1):1–12.

- Fuller J (1995). Conditioning Stress Development and Factors that Influence the Prong Test. Research Paper FPL-RP-537, Forest Products Laboratory, Madison, WI, USA.

Slitting Method

- Prime MB (1999). Residual Stress Measurement by Successive Extension of a Slot: The Crack Compliance Method. *Applied Mechanics Reviews*, 52(2):75–96.

- Cheng W, Finnie I (2007). *Residual Stress Measurement and the Slitting Method*, Springer, New York.

- Lee MJ, Hill MR (2007). Intralaboratory Repeatability of Residual Stress Determined by the Slitting Method, *Experimental Mechanics*, 47(6):745–752.

Ring-Core Method

- Kiel S (1992). Experimental Determination of Residual Stresses with the Ring-Core Method and an On-Line Measuring System. *Experimental Techniques*, 16(5):17–24.

- Ajovalasit A, Petrucci G, Zuccarello B (1996). Determination of Non-uniform Residual Stresses using the Ring-Core Method. *Journal of Engineering Materials and Technology*, 118(2):224–228.

- Barsanti M, Beghini M, Santus C, Benincasa A, Bertelli L (2014). Integral Method Coefficients and Regularization Procedure for the Ring-Core Residual Stress Measurement Technique, *Advanced Materials Research*, 996(1):331–336.

Hole Drilling Method

- ASTM (2013). Standard Test Method for Determining Residual Stresses by the Hole-Drilling Strain-Gage Method, Standard Test Method E837-13, ASTM International, West Conshohocken, PA.

- Grant PV, Lord JD, Whitehead PS (2002). The Measurement of Residual Stresses by the Incremental Hole Drilling Technique. Measurement Good Practice Guide, No. 53. National Physical Laboratory, UK.

- Measurements Group (2001). Measurement of Residual Stresses by Hole-Drilling Strain Gage Method. Tech Note TN-503-6, Vishay Measurements Group, Raleigh, NC.

- Ajovalasit A, Scafidi M, Zuccarello B, Beghini M, Bertini L, Santus C, Valentini E, Benincasa A, Bertelli L (2010). The Hole-Drilling Strain Gauge Method for the Measurement of Uniform or Non-Uniform Residual Stresses, *AIAS Residual Stress Working Group, (TR01)*, Florence, Italy.

Deep-Hole Drilling Method

- Leggatt RH, Smith DJ, Smith S.D, Faure F (1996). Development and Experimental Validation of the Deep Hole Method for Residual Stress Measurement. *The Journal of Strain Analysis for Engineering Design*, 31(3):177–186.

- DeWald AT, Hill MR (2003). Improved Data Reduction for the Deep-Hole Method of Residual Stress Measurement. *The Journal of Strain Analysis for Engineering Design*, 38(1):65–78.

Layer Removal

- Treuting RG, Read WT (1951). A Mechanical Determination of Biaxial Residual Stress in Sheet Materials. *Journal of Applied Physics*, 22(2):130–134.

- Sachs G, Espey G (1941). The Measurement of Residual Stresses in Metal. *The Iron Age*, pp. 63–71, September 18.

- Stoney GG (1909). The Tension of Metallic Films Deposited by Electrolysis. *Proc. of the Royal Society of London, Series A*, 82(553):172–175.

Contour Method

- Prime MB, DeWald AT (2013). The Contour Method. Chapter 5 in *Practical Residual Stress Measurement Methods*, Schajer GS (Ed.), Wiley, West Sussex, UK.

- Prime MB (2001). Cross-Sectional Mapping of Residual Stresses by Measuring the Surface Contour After a Cut. *Journal of Engineering Materials and Technology*, 123(2):162–168.

- Kelleher J, Prime MB, Buttle D, Mummery PM, Webster PJ, Shackleton J, Withers PJ (2003). The Measurement of Residual Stress in Railway Rails by Diffraction and Other Methods. *Journal of Neutron Research*, 11(4):187–193.

Sectioning

- Shadley JR, Rybicki EF, Shealy WS (1987). Application Guidelines for the Parting Out Step in a Through Thickness Residual Stress Measurement Procedure. *Strain*, 23(4):157–166.

- Tebedge N, Alpsten G, Tall L (1973). Residual-stress Measurement by the Sectioning Method. *Experimental Mechanics*, 13(2):88–96.

Microscopic Scale Measurements

- Winiarski B, Withers PJ (2015). Novel Implementations of Relaxation Methods for Measuring Residual Stresses at the Micron Scale. *The Journal of Strain Analysis for Engineering Design*, 50(7):412–425.

- Winiarski B, Withers PJ (2012). Micron-Scale Residual Stress Measurement by Micro-Hole Drilling and Digital Image Correlation, *Experimental Mechanics*, 52(4):417–428.

- Song X, Yeap KB, Zhu J, Belnoue J, Sebastiani M, Bemporad E, Zeng K, Korsunsky AM (2012). Residual Stress Measurement in Thin Films at Sub-Micron Scale Using Focused Ion Beam Milling and Imaging, *Thin Solid Films*, 520(6):2073–2076.

X-Ray Diffraction

- SAE (2003). Residual Stress Measurement by X-ray Diffraction. SAE J784, *Society of Automotive Engineers Handbook Supplement*, Warrendale, PA.

- Fitzpatrick ME, Fry AT, Holdway P, Kandil FA, Shackleton J, Suominen L (2005). Determination of Residual Stresses by X-ray Diffraction, Measurement Good Practice Guide No. 52 Issue 2, National Physical Laboratory, Teddington, Middlesex, UK.

- Noyan IC, Cohen JB (1987). *Residual Stress—Measurement by Diffraction and Interpretation, Materials Research and Engineering*, Springer-Verlag, New York.

Synchrotron Diffraction

- Withers PJ (2013). Synchrotron X-ray Diffraction, Chapter 7 in *Practical Residual Stress Measurement Methods*, Schajer GS (Ed.), Wiley, West Sussex, UK.

- Reimers W, Pyzalla A, Broda M, Brusch G, Dantz D, Schmackers T, Liss K-D, Tschentscher T (1999). The Use of High-Energy Synchrotron Diffraction for Residual Stress Analyses. *Journal of Materials Science Letters*, 18(7):581–583.

- Steuwer A, Santisteban JR, Turski M, Withers PJ, Buslaps T (2004). High-resolution Strain Mapping in Bulk Samples using Full-profile Analysis of Energy Dispersive Synchrotron X-ray Diffraction Data. *Journal of Applied Crystallography*, 37(6):883–889.

Neutron Diffraction

- Fitzpatrick ME, Lodini A (Eds.) (2003). *Analysis of Residual Stress by Diffraction using Neutron and Synchrotron Radiation*, Taylor and Francis, London, UK.

- Holden TM (2013). Neutron Diffraction, Chapter 8 in *Practical Residual Stress Measurement Methods*, Schajer GS (Ed.), Wiley, West Sussex, UK.

Magnetic Methods

- Jiles DC (1988). Review of Magnetic Methods for Nondestructive Evaluation. *NDT International*, 21(5):311–319.

- Buttle DJ, Moorthy V, Shaw B (2006). Determination of Residual Stresses by Magnetic Methods, Measurement Good Practice Guide, No. 88. National Physical Laboratory, Teddington, Middlesex, UK.

- Buttle DJ (2013). Magnetic Methods. Chapter 9 in *Practical Residual Stress Measurement Methods*, Schajer GS (Ed.), Wiley, West Sussex, UK.

Ultrasonic

- Leon-Salamanca T, Bray DE (1996). Residual Stress Measurement in Steel Plates and Welds using Critically Refracted Longitudinal (LCR) Waves. *Research in Nondestructive Evaluation*, 7(4):169–184.

- Santos AA, Bray, DE (2000). Application of Longitudinal Critically Refracted Waves to Evaluate Stresses in Railroad Wheels. *Topics on Nondestructive Testing*, 5, American Society for Nondestructive Testing.

Thermoelastic

- Wong, AK, Dunn SA, Sparrow JG (1988). Residual Stress Measurement by Means of the Thermoelastic Effect. *Nature*, 332(6165):613–615.

Photoelastic

- Hetenyi M (Ed.) (1950). *Handbook of Experimental Stress Analysis*, Wiley, New York.

CHAPTER 3

Hole-Drilling Method Concept and Development

"The new test method ... is done by drilling a hole, which, however, is so small that the part can be used again."

Josef Mathar (1934) in *Determination of Initial Stresses by Measuring the Deformations Around Drilled Holes.*

3.1 INTRODUCTION

The Hole-Drilling Method has an extensive history dating back to the 1930s. It has developed into a modern residual stress measurement technique that uses state-of-art technology to enable non-specialists to achieve accurate results consistently and reliably. This state of development has been achieved through the accumulation of many individual advances by numerous workers over the years. This chapter gives an overview of the operational concept and the historical development of the Hole-Drilling Method from the early explorations to contemporary detailed knowledge.

3.2 CONCEPT

The Hole-Drilling Method involves drilling a small hole in the material specimen at the desired test location. The removal of stressed material within the hole causes a stress redistribution in the material surrounding the hole. This stress redistribution, or "relaxation," creates localized elastic deformations in that material. Figure 3.1 schematically illustrates the outward deformations around a hole drilled into material with tensile residual stresses. There is also a small local surface rise caused by the action of Poisson's ratio; the reverse happens in the presence of compressive stresses. In practical measurements, the relaxation deformations of the material around the hole are measured using strain gauges or by optical techniques. Specialized calculation procedures are then used to evaluate the residual stresses that originally existed within the hole from the measured deformations.

The closely related Ring-Core Method reverses the geometry of the Hole-Drilling Method by placing the measurement area in the middle and making the "hole" in the form of a surrounding annular groove. Figure 3.2 compares the geometry of the Hole-Drilling and

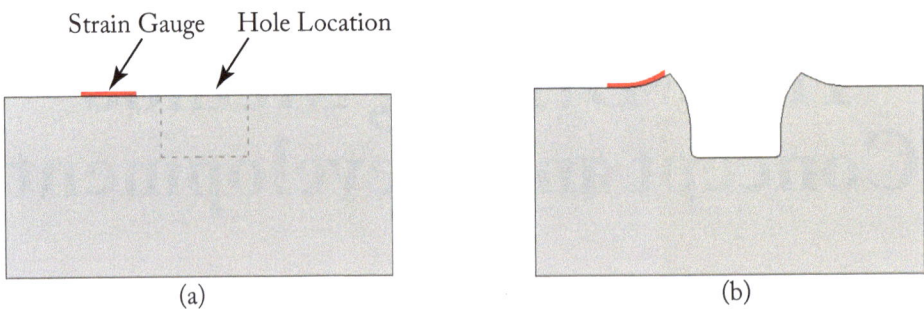

Figure 3.1: Schematic cross-sections around a hole drilled into tensile residual stresses. (a) Before hole drilling and (b) after hole drilling.

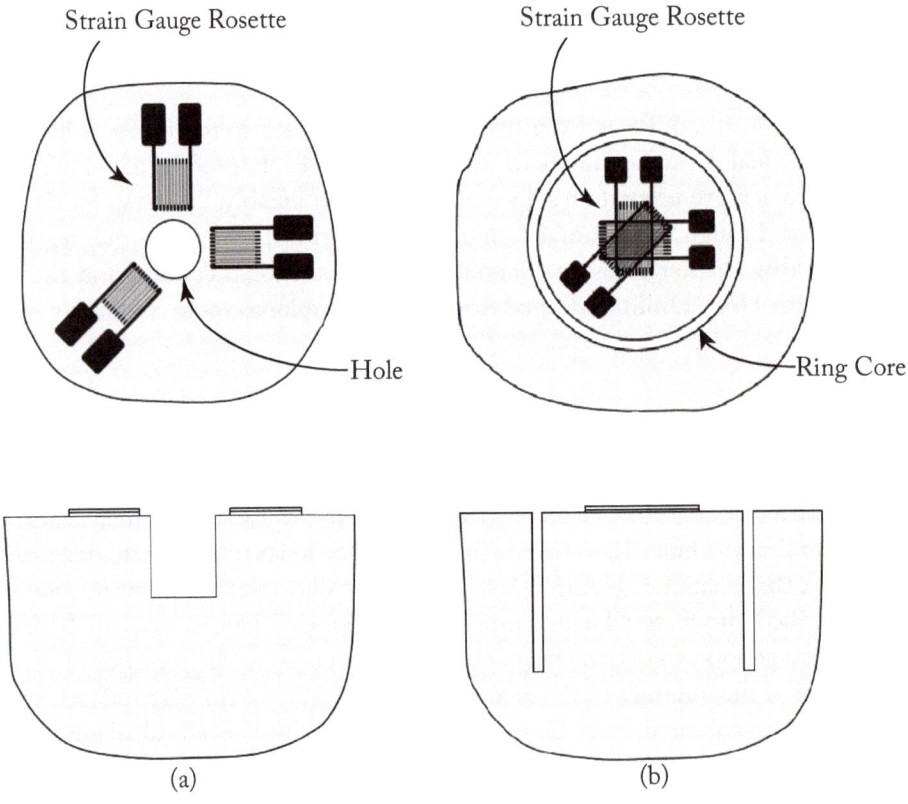

Figure 3.2: Residual stress measurement methods. (a) Hole drilling and (b) ring-core.

Ring-Core methods. The two methods use identical residual stress evaluation procedures, and differ only in the numerical constants needed. The Ring-Core Method has the advantage of producing larger relieved strains and has superior capability to measure very large residual stresses close to the material yield stress. However, the Hole-Drilling Method is the more commonly used procedure because of its much greater ease of use and lesser specimen damage.

3.3   MATHAR'S FOUNDATIONAL WORK

The Hole-Drilling Method derives from the pioneering work of Josef Mathar in the early 1930s. His two papers on hole-drilling, the first in German in 1932 and the second in English in 1934, lay the foundations of the Hole-Drilling Method. In his writings he also explicitly anticipates both the Ring-Core Method and the Deep-Hole Method for making geological stress measurements, so he may additionally be considered as a spiritual ancestor of these two methods as well.

Figure 3.3 illustrates the device that Mathar developed for doing hole-drilling residual stress measurements. It used a twist drill to cut a stepped hole, 12 mm in diameter, and a mechanical extensometer to measure the resulting deformations around the hole. Although basic by modern standards, the drilling/measuring device is remarkable for its identification and implementation of most of the fundamental features needed of such a device, thereby providing the prototype for subsequent drilling/measuring devices. Figure 4.6 in Chapter 4 illustrates some modern hole drilling/measuring devices that incorporate updated implementations of the early functionalities.

Mathar's pioneering work addressed and gave initial solutions to the three main procedural aspects of hole-drilling residual stress measurements:

1. drilling the hole,

2. measuring the resulting deformations around the hole, and

3. computing the residual stresses from the measured deformations.

These three procedural aspects remain the main issues in modern practice. The following sections describe the subsequent developments of each of them. These developments have involved a mixture of both technical and conceptual advances.

3.4   HOLE DRILLING

The procedure for cutting the hole required for hole-drilling measurements has greatly evolved since the early days. In his original work, Mathar used a stepped tool comprising a short 5-mm diameter pilot drill leading a 12-mm diameter endmill style cutter for his residual stress measurements. He chose this cutting tool design to address concerns for accurate hole alignment and for minimal machining stress introduction. These issues remain the main concerns in modern hole-drilling practice.

Figure 3.3: Hole-drilling/measurement device developed by Mathar. q = tensometer indicator, u = mounting arm, v = depth gauge, z = drilling machine, f = clamping frame, and t = sleeve (Mathar (1934)).

Following Mathar's lead, early researchers typically used modified twist-drills or milling cutters for hole drilling. Figure 3.4 shows examples of some of the designs used. Up to the 1970s, drill rotation speeds were generally quite low, in the 200–2000 rpm range, with drive provided either manually or using a power drill. Figure 3.5 shows examples of such drilling devices. Interestingly, the modern RS-200 drilling guide, which has developed substantially since its original introduction, still includes among its optional components a low-speed drilling accessory as a vestigial remnant of former practice.

In the 1970s, Flaman and coworkers popularized the use of high-speed dental burs for hole-drilling residual stress measurements. They showed that the use of high rotation speeds, reported as up to 400,000 rpm, could minimize the introduction of machining stresses during the hole cutting. The desire to reduce such stresses had been a major motivation for the early modified cutter designs shown in Figure 3.4. Miniature air turbines of the type commonly used for dental drills became standard for hole cutting, together with the use of tungsten carbide dental burs as the cutters, most typically the inverted cone design shown in Figure 4.8 in Chapter 4.

Practical on-load rotation speeds of air-turbines are commonly in the range 50,000–150,000 rpm. In recent years, high-speed electric drives have become a popular alternative choice

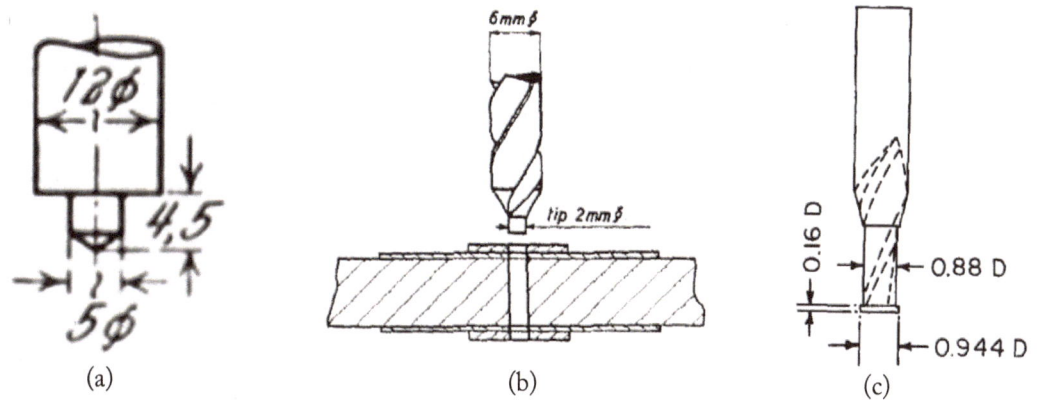

Figure 3.4: Examples of modified drilling cutters used by early researchers. (a) Mathar (1934), (b) Boiten and ten Cate (1953), and (c) Rendler and Vigness (1966).

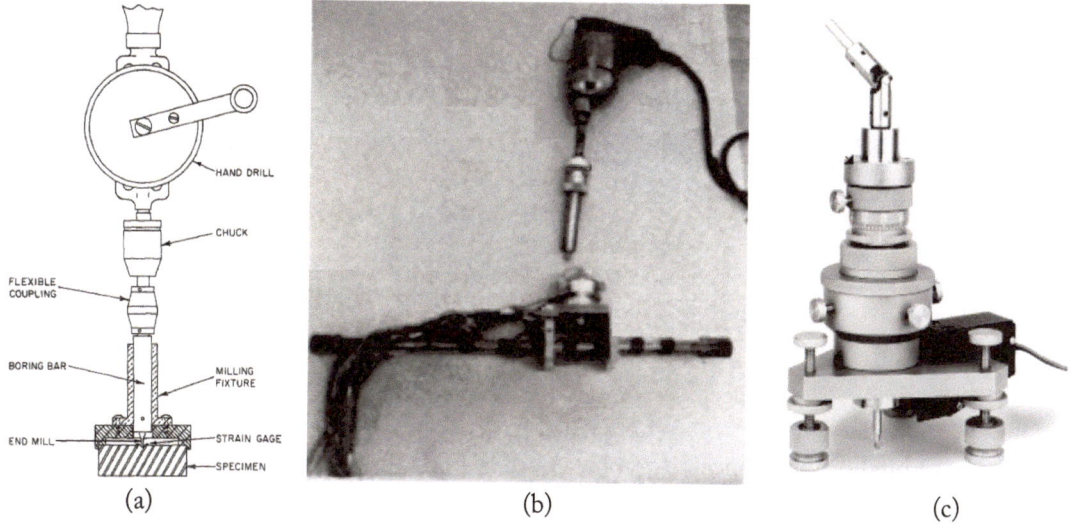

Figure 3.5: Examples of low-speed hole drilling devices. (a) Rendler and Vigness (1966), (b) Flaman (1982), and (c) Old version RS-200 (reproduced by permission of Micro-Measurements, a Vishay Precision Group brand).

because they avoid the inconvenience of having to provide a pressurized air supply. Steinzig et al. (2014) demonstrated that their lower rotation speeds in the >10,000 rpm range are still effective for minimizing cutting stresses. These drives typically have industrial style collets as tool holders. This feature widens the range of choices for cutting tools to include miniature milling cutters with titanium or chromium nitride and other coatings. Such tools enable effective cutting of hard materials that previously were very difficult to work with using conventional cutters.

Orbital drilling was another technique introduced by Flaman. This process involves off-setting the drill motor within its housing and then rotating the whole housing within its holder to produce an orbiting path. Figure 3.6 illustrates the process. One advantage of orbital cutting is that it moves the main cutting action from the end face of the cutter to the more effective side cutting edges. This feature also allows the chips produced during the cutting to escape freely through the open area opposite the drill offset. In contrast, the chips produced during traditional plunge cutting must make their way from the bottom of the hole, along the narrow drill flutes eventually to reach the top. On their journey, these chips rub along the hole surface, causing susceptibility to introduction of machining stresses. Other advantages include the reduction in the drilling force normally associated with pressing the stationary drill center into the specimen during conventional axial drilling.

Figure 3.6: Orbital drilling concept (graphic courtesy of Stresscraft Ltd.).

The name "Hole-*Drilling* Method" must be considered as a generic rather than specific name because some hole production processes do not involve drilling. An interesting alternative approach developed by Beaney and Procter in the 1970s is to use a jet of abrasive particles from a fine nozzle to create a hole by controlled mechanical erosion. This hole production method, called "air-abrasive," is particularly effective when working with very hard materials. Such materials are very challenging to drill using conventional cutters, even by those with hard coatings. A further feature of the air-abrasive process is that it does not create significant machining stresses. However, the shape of the hole made by air-abrasion does not have the well-defined cylindrical shape of the type produced by a conventional cutter, although orbiting the nozzle to create an action similar to that in Figure 3.6 substantially helps. Because of the imperfect hole shape, the air-abrasive method is suitable only for measurements of uniform residual stresses.

An alternative "hole-drilling" procedure is by Electro-Discharge Machining (EDM) using an electrode with shape corresponding to the desired hole cross-section. The EDM technique has similar advantages to the air-abrasive technique in that it can work with very hard materials, but great care is required to minimize the thickness of the "re-cast layer" formed during the process. Electrochemical Machining (ECM) is another possible method of forming a blind hole that can proceed in an entirely stress-free manner, but control of electrolyte in close proximity to the strain gauge is problematic. Air-abrasive, EDM and ECM methods all require rather specialized equipment, so have only specialized use.

Contemporary hole-drilling equipment and practices have developed and continue to develop in many details from their historical predecessors. Figure 3.7 illustrates some contemporary hole-drilling equipment. The use of such devices has greatly enhanced the ease and range of use of the hole-drilling method.

(a)

(b)

Figure 3.7: Modern hole-drilling devices. (a) Vishay RS-200 with air turbine (reproduced by permission of Micro-Measurements, a Vishay Precision Group brand) and (b) SINT drilling device (reproduced by permission of SINT Technology s.r.l.).

3.5 DEFORMATION MEASUREMENTS

From an early stage, the mechanical extensometer used by Mathar was recognized as a major factor limiting the accuracy and reliability of hole-drilling residual stress measurements. The development of strain gauges in the 1940s enabled substantial improvements to be achieved in deformation measurement quality. In 1950, Soete introduced the use of strain gauges for hole-drilling measurements, greatly improving measurement accuracy and reliability, and allowing smaller holes to be used. At that time, strain gauges were still in their infancy and their use required specific consideration to fit the needs of each intended application. Modern process-ing techniques for thermal stabilization had not yet been introduced, so thermal stability of the strain gauges was a particular concern. This concern was addressed by using multiple strain gauges in half or full bridge configurations. Figure 3.8 shows some early strain gauge rosette ar-rangements, designed to give a three-axis measurements with full thermal compensation. These designs are interesting from an historical point of view for the insight they give into the concerns and challenges facing early strain gauge practitioners.

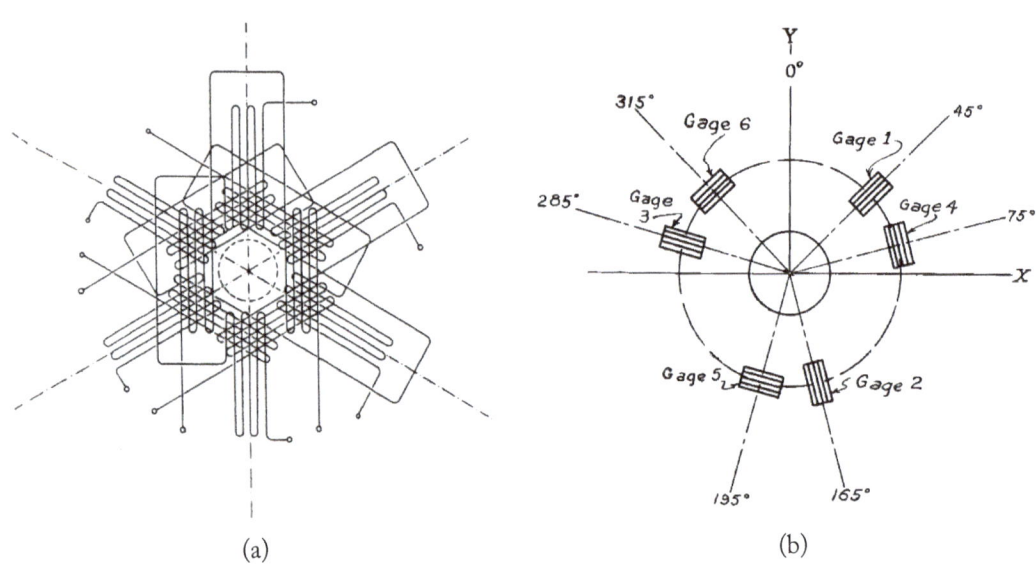

Figure 3.8: Early hole-drilling strain gauge rosette designs with thermal compensation. (a) Ri-parbelli (1950) and (b) Riparbelli (1958).

Several of the early researchers, notably Soete and Vancrombruge, Riparbelli; Boiten and ten Cate; Kelsey; Rendler and Vigness; and Bathgate, pioneered the use of strain gauges for hole-drilling residual stress measurements. In general, they adapted general-purpose strain gauges for hole-drilling use. This involved very careful placement of the strain gauges to achieve an accurate, symmetrical arrangement around the intended hole location.

The work of Rendler and Vigness (1966) can be considered the foundation of the modern strain gauge hole-drilling practice through their introduction of preassembled rosettes using a standardized strain gauge geometry. The custom design and precision construction of the rosettes provide much superior performance compared with the early hand-assembled strain gauges. In addition, Rendler and Vigness recognized that the measured strain response depends only on the relative sizes of the rosette components and the hole, independent of their overall dimensions. Thus, all rosette and hole dimensions can be scaled up or down as desired to suit experimental needs. These developments advanced the hole-drilling method into a practical procedure that could be used by general practitioners.

The work of Rendler and Vigness provided the basis for the establishment of ASTM Standard Test Method E837 in 1981, updated and expanded several times since then. Figure 3.9 shows the standardized strain gauge rosette designs specified by E837. The geometry of the "Type A" pattern corresponds exactly with the original Rendler and Vigness design and is available in three different sizes. The 1/8" and 1/16" sizes come directly from Rendler and Vigness's work, the smaller 1/32" size was added subsequently.

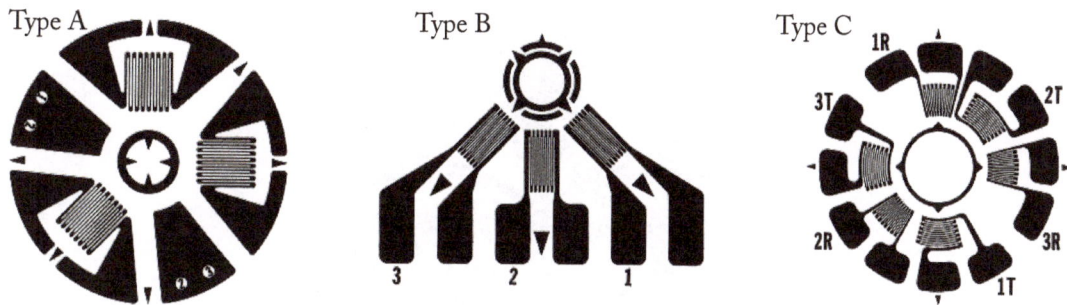

Figure 3.9: Standardized hole-drilling strain gauge rosettes (reproduced by permission of Micro-Measurements, a Vishay Precision Group brand).

The "Type B" rosette in Figure 3.9 is a variant geometry, designed for use near a boundary or obstacle. The "Type C" rosette is a more specialized design intended for use on low thermal conductivity materials or where the local residual strains are very low. This rosette pattern achieves in modern format the thermal compensation objective anticipated in the 1958 Riparbelli rosette design shown in Figure 3.8b.

Since their introduction in the 1950s, strain gauges have been the standard sensor used for measuring surface deformations when making hole-drilling residual stress measurements. There are many good reasons for this; they are convenient to use, they give reliable and accurate measurements and the supporting instrumentation is widely available at moderate cost. These factors have combined with the conceptual simplicity and generality of the hole-drilling process to make the Hole-Drilling Method one of the most commonly used residual stress measurement methods. However, this is not to imply that strain gauges are the only sensor choice, nor universally the ideal choice.

Starting in the 1980s, full-field optical methods based on interferometry were introduced for hole-drilling residual stress measurements. Such methods are attractive because they display a detailed map of the deformations around the drilled hole and because they avoid the need to attach strain gauges on the specimen surface.

Initial optical developments were made using interferometric Moiré and holographic techniques. In the 1990s, a variant technique called Electronic Speckle Pattern Interferometry (ESPI) also started to be applied to hole-drilling measurements. The ESPI technique uses a digital camera to record the optical images and is attractive because it eliminates the need for a thermoplastic or similar recording device used by the holographic method. Figure 3.10 shows an example of an ESPI measurement. The light and dark lines are called "fringes" and the intervals between them represent incremental surface displacements of a fraction of the wavelength of the light used, the exact fraction depending on the interferometer geometry. All interferometric techniques make measurements at the scale of the wavelength of light, 400–700 nm in the visible range, so they are highly sensitive and can detect very small surface displacements.

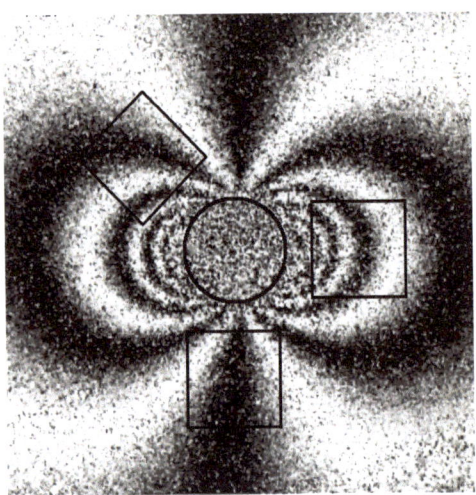

Figure 3.10: A typical ESPI fringe pattern from a hole-drilling measurement. The circle and squares, respectively, show the hole and equivalent strain gauge locations.

Starting in the 2000s, the popular Digital Image Correlation (DIC) technique started to be applied to hole-drilling residual stress measurements. It has the advantage of being more tolerant of field conditions than the interferometric methods, although for conventional size holes in the 1–5 mm range, it is less sensitive. However, with care, satisfactory results can still be achieved.

DIC differs in an important fundamental way from the interferometric methods. The latter are scale-dependent because their measurements are produced relative to the wavelength of the light used. Thus, they work best with macro-scale holes, typically with diameter greater than about 0.5 mm. In contrast, DIC measurements are scale-independent because their sensitivity is controlled solely by the pixel dimensions and the feature density of the measured images, but not on the physical dimensions that the images represent. Thus, if a hole occupies a given fraction of an image, the same relative measurement resolution will be achieved independent of whether the hole is physically 100 mm, 1 mm, or 1 μm in diameter. Practical DIC hole-drilling measurements have been done at both ends of this range, the biggest holes produced by cutting cores out of large concrete members, and the smallest made using a Focused Ion Beam within a Scanning Electron Microscope (FIB-SEM). The concrete example demonstrates measurements at a scale too big for conventional strain gauge use, while the microscopic example demonstrates measurements at a much smaller scale than possible with traditional strain gauges. Figure 3.11 illustrates a DIC hole-drilling measurement done within a scanning electron microscope. The hole diameter is approximately 1 μm. The surrounding dot pattern is deliberately applied to provide data for the DIC evaluation. The surface deformations in the x and y directions caused by hole-drilling can then be determined throughout that area.

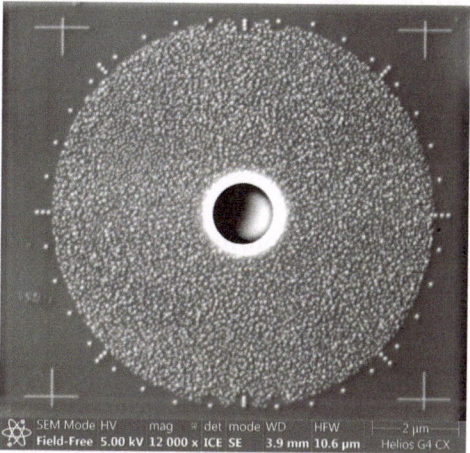

Figure 3.11: DIC hole-drilling measurement done using a FIB-SEM (image courtesy of Dr. B. Winiarski, Manchester University, UK).

3.6   RING-CORE METHOD

The Ring-Core Method, schematically illustrated in Figure 3.2b, is closely analogous to the Hole-Drilling Method, but with the locations of the hole and the measurement area interchanged. Instead of having a hole at the center and measurements around the outside, the measurements are at the center and the "hole" becomes an annular groove around the outside. The Ring-Core Method is an evolution of the excision method described in 1946 by Meriam et al. They measured residual stresses in welded steel plates by attaching strain gauge rosettes and then entirely cutting out the local stressed material by drilling a series of overlapping holes along a surrounding circular path. The modern form of the Ring-Core Method using an annular groove was introduced by Milbradt in 1951, with subsequent developments by Beaney and Proctor, Keil and several others. In many ways, the development of the Ring-Core Method has proceeded in parallel with that of the Hole-Drilling Method, for example, optical measurements have similarly been applied to Ring-Core measurements, as illustrated in Figure 3.12. The residual stress computational procedures required for both methods are identical, differing only in the numerical values of the calibration constants. Thus, there is substantial cross-fertilization and transfer of ideas between practitioners of the two methods.

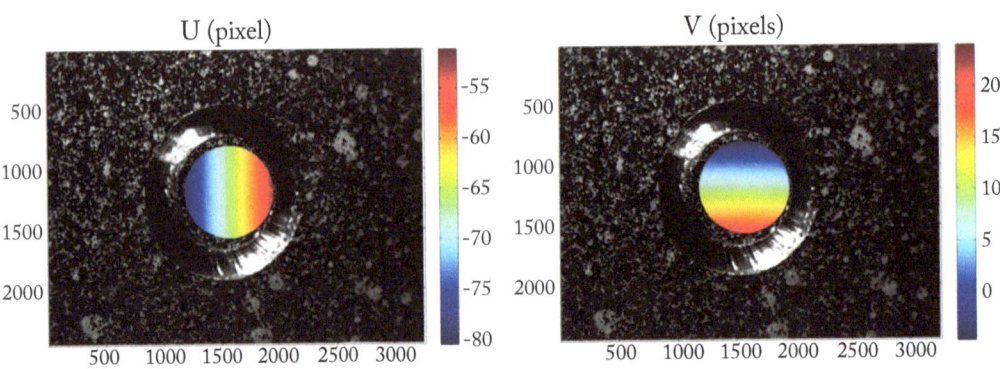

Figure 3.12: Application of DIC for Ring-Core measurements (from Hu et al. (2013)). (a) Horizontal displacements and (b) vertical displacements.

The experimental procedure of the Ring-Core Method tends to be less convenient than that of the Hole-Drilling Method and also the ring cutting process creates much greater specimen damage than hole drilling. These characteristics cause the Ring-Core Method to be very much less commonly used, likely by over an order of magnitude. However, the Ring-Core Method can still be a practical choice in many cases because of its larger relieved strains and it superior capability to measure very large residual stresses close to the material yield stress.

3.7   DEEP-HOLE DRILLING

Deep-Hole Drilling is a further variant approach that combines several elements of the Hole-Drilling and Ring-Core Methods. The method involves drilling a deep hole into the test material and then measuring the change in diameter as the surrounding material is overcored. The technique was suggested by Mathar in 1934 as a possible way to measure geological stresses within large rock masses, but practical implementations came much later, notably by Leeman (1968) and Merrill (1967). The technique was also applied to the measurement of residual stresses in large metal components such as castings by Beaney and very extensively further developed by Smith and coworkers. Figure 3.13 schematically illustrates a typical large-metal application.

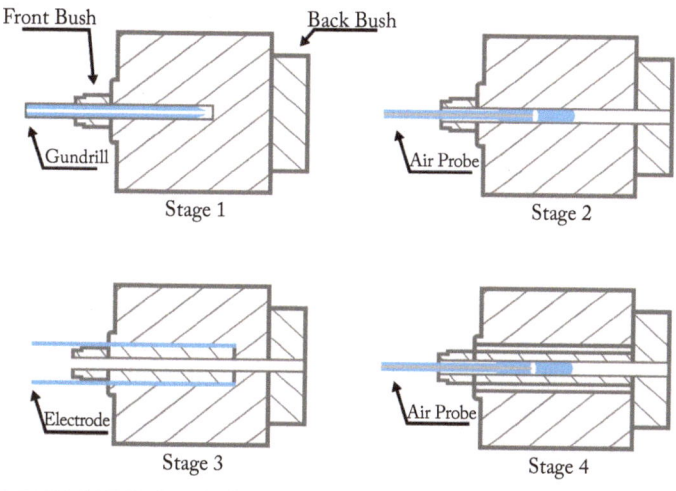

Figure 3.13: Deep-hole measurement in a large metal component. (1) Attach front and back bushes and gun-drill a pilot hole, (2) measure the diameter along the length of the pilot hole, (3) overcore the pilot hole, and (4) re-measure the diameter along the length of the pilot hole (diagram courtesy of Veqter Ltd.).

A fundamental distinguishing feature of the Deep-Hole Method is that the deformation measurements are made within the interior of the hole rather than at the surface. This is done to move the location of stress sensitivity into the deep interior. In contrast, hole-drilling and ring-coring measurements focus on near-surface stresses because their measurements are at the surface. For metal components, deep-hole measurements can be made to about 750 mm deep, while for geological applications, measurement depths over 1 km are routinely achieved.

3.8 RESIDUAL STRESS COMPUTATIONS

For the Ring-Core Method, calculation of the residual stresses corresponding to the measured strains is fairly straightforward because complete strain relief occurs within the central core and so the standard equations of linear elasticity may be directly applied. However, for the Hole-Drilling Method, only a fraction of the strains within the material surrounding the drilled hole are relieved, so the residual stress calculation also requires knowledge of the specific numerical value of that fraction. In practice, the needed data take the form of calibration constants provided either in numerical or graphical form.

Historically, two approaches have been taken to determine hole-drilling calibrations, theoretical and experimental. In his pioneering 1934 work, Mathar initiated the use of both approaches. On the theoretical side, he used the classical Kirsch solution for the stress distribution around a through hole in a thin stressed plate to determine the relationship between hole diameter change and the in-plane stresses. On the experimental side, he also measured diameter change vs. hole depth for known in-plane stresses for the case of blind hole drilling in a thick material. He noted that the experimental results asymptotically approach the theoretical solution when the hole depth reaches approximately one hole diameter.

Both the theoretical and experimental approaches were taken up by subsequent researchers. The greater accuracy provided by the advent of strain gauge use in the 1940s and 1950s made it possible to go beyond the simple evaluation of the bulk residual stresses at the hole location to the determination of the profile of the residual stresses through the hole depth. Since the Kirsch analytical solution only applies to stresses that are uniform within a through-hole in a thin plate, calibrations for non-uniform stresses within a blind hole in thick plate could at that time only be done experimentally. Thus, computational procedures had to be chosen to use only experimental calibration data.

Kelsey (1956) described a stress profile computation method that later become known as the Differential Strain Method. The method is based on the assumption that the change in strain measured at a given hole depth depends on the stresses existing at that depth. Thus, the residual stress profile can be determined by comparing point-by-point the slopes of the strain vs. depth profiles of the test measurements with experimental calibration data for a known uniform stress state. This method works reasonably well for residual stress profiles that do not greatly differ from uniform, however its accuracy significantly deteriorates when used with significantly non-uniform residual stress profiles. The reason is that the evolution of the measured strain response additionally depends on the hole geometry change caused by its increasing depth. Thus, the change in relieved strain at any stage combines the effects of hole enlargement on the previously exposed near-surface stresses as well as those on the newly exposed interior stresses.

The Average Stress Method introduced by Nickola in 1986 sought to address the concerns with the Differential Strain Method by recognizing that the strain response at a given hole depth combines the effects of all the stresses at the various depths within the hole depth. A running average stress at each hole depth is computed using the calibration data for the uniform stress case. The stress within each hole depth increment is then evaluated by determining the stress values that combine to give the computed sequence of running averages. In practice, the strain response is highly nonlinear, with much greater sensitivity to near-surface stresses than to interior stresses. Thus, the use of a simple average stress value underweights the effects of the near-surface stresses and overweights those of the interior stresses. Thus, the Average Stress Method also has limited usefulness and is also suitable only for near-uniform residual stress profiles.

The growth in the practical use of the finite element method in the 1970s enabled the development of residual stress computation methods that could break through the limitations imposed by dependence on experimental calibrations. During the 1970s and 1980s, advances were made in three important areas:

1. accuracy improvement of hole-drilling calibration data,

2. formulation of calibration data not measurable experimentally, and

3. creation of calibration data for additional applications.

The early works of Bijak-Zochowski (1978), Beaney and Procter (1974), Schajer (1981) were aimed at improving calibration accuracy and consistency and established the finite element method as a practical method for evaluating hole-drilling calibration data. An important development was the introduction of the Integral Method for calculating the residual stress profile through the depth of the drilled hole. Several researchers contributed to this initial development, notably Bijak-Zochowski, Flavenot and Lu, and Schajer (1988). The significant feature of the Integral Method is that it correctly accounts for the contributions of all stresses within the hole depth to the measured strain response, thereby avoiding the limitations of the Differential Strain and Average Stress methods. The use of finite element calculations was an essential prerequisite to the development and use of the Integral Method because the needed calibration data are not measureable experimentally. Since 1999, the ASTM Standard Test Method E837 has specified the use of the Integral Method for residual stress profile evaluations.

The flexibility of the finite element method has enabled substantial computational developments to be made in many further areas, notably for corrections for various experimental artifacts that can occur during the course of practical measurements. One such artifact occurs when the drilled hole is not exactly at the geometric center of the strain gauge rosette. The resulting eccentricity causes a systematic shift in the measured strains, thereby distorting the corresponding computed residual stresses. In the late 1970s, Sandifer and Bowie (1978) and Ajovalasit (1979) introduced computational approaches for correcting the effect providing that the size and direction of the hole eccentricity are accurately known. However, the opportunity to correct for hole eccentricity should not be allowed to grant a tolerance of such errors. Certainly it is always the best strategy is to seek to refine the experimental technique used so as to minimize the occurrence and size of all preventable errors.

Another significant artifact occurs when the Hole-Drilling Method is used to measure large residual stresses close to the material yield stress. The drilling of the hole creates a stress concentration in the adjacent material, which causes local yielding. The resulting material plasticity adds to the relieved strains, causing them to be larger than they would be if only elastic deformations had occurred. Consequently, the residual stresses evaluated from the measured strain reliefs are overestimated, often suggesting residual stresses larger than the material yield stress. This artifact can also be ameliorated using finite-element based compensation methods. Starting in the 1990s, Beghini and coworkers have led the initiative to develop approaches to compensate for material plasticity and to allow hole-drilling measurements to be made for residual stresses close to yield stress. Without such corrections, the Hole-Drilling Method can be used only for measuring residual stresses up to about 60% of the yield stress, or to 80% if the hole depth is limited to about half its diameter, as specified in the 2013 revision of ASTM E837. The greater stress capacity at smaller hole depths occurs because the material below the cut hole reinforces the surrounding material and reduces the stress concentration effect of the hole.

The advent of optical methods for measuring the deformations about the drilled hole created new computational challenges because of the large quantity but relatively low quality of the available data. Early computational procedures tended to follow strain gauge practice. Displacement data were selected from a small number of hypothetical strain gauge locations within the overall measurement area and then used with the established strain gauge equations to determine stresses. In some cases, such as shown in Figure 3.11, the specific areas corresponding to standard strain gauge rosette designs were chosen so that the associated calibration data could be used.

Later computational approaches sought to take advantage of the large number of data available from optical measurements, often running into the millions. These data can be averaged using various techniques so as to reduce the effect of measurement noise and to improve overall measurement quality. The data can also be analyzed to derive quantities other than the three in-plane stresses. Notably, artifacts due to thermal effects, bulk displacements, rotations and scale changes can be identified and separated out. The latter separations are particularly important when doing microscopic measurements because small fluctuations in image settings are inevitable and will cause significant deviations if not compensated.

Development of computational procedures remains an active and productive area of development of the Hole-Drilling Method, stimulated in large part by the ever improving accessibility to sophisticated finite element software and by the interesting opportunities presented by the rich data available from optical measurements. The challenge is to create straightforward procedures that are compact and practical and can be used by a wide cross-section of practitioners.

3.9   CONCLUDING REMARKS

The Hole-Drilling Method for measuring residual stresses has grown and developed substantially since the pioneering work of Mathar. The technique has become well established, with its own ASTM Standard Test Procedure. The Hole-Drilling Method continues to advance actively in all three of its main aspects: hole-drilling, deformation measurement and computational methods. Recent work has concentrated on the use of full-field optical techniques to measure the deformations around a drilled hole. These developments have greatly expanded the scope of hole-drilling residual stress measurements, notably by providing a very rich source of available data suitable for further analysis and evaluation.

Mathar's foundational idea to use the deformations around a drilled hole to evaluate material residual stresses has proven to be a very fertile one. Nearly a century of active research and application has followed his pioneering work. Unfortunately, the early death of Mathar prevented him from witnessing the growth and development of his concept. Had he lived on, he would certainly have had good reason to be proud of his innovation and the substantial further activity that it initiated. The Hole-Drilling Method continues as a very active field of research and application, with a flow of imaginative conceptual and procedural advances that shows no signs of abating.

3.10 FURTHER READING

Hole Drilling Concept

- Mathar J (1934). Determination of Initial Stresses by Measuring the Deformation Around Drilled Holes. *Transactions of the American Society of Mechanical Engineers*, 56(4):249–254.

- ASTM (2013). Determining Residual Stresses by the Hole-drilling Strain-gage Method. *Standard Test Method E837-13*, American Society for Testing and Materials, West Conshohocken, PA.

- Grant PV, Lord JD, Whitehead PS (2002). The Measurement of Residual Stresses by the Incremental Hole Drilling Technique. *Measurement Good Practice Guide*, No. 53, National Physical Laboratory, Teddington, UK.

- Vishay Measurements Group, Inc. (1993). Measurement of Residual Stresses by the Hole-drilling Strain-gage Method. *Tech Note TN-503-6*, Vishay Measurements Group, Inc., Raleigh, NC. 16pp.

- Schajer GS and Whitehead P (2013). Hole-drilling and Ring Core Methods. Chapter 2 in *Practical Residual Stress Measurement Methods*, G. Schajer (Ed.), Wiley, Chichester, UK.

- Ajovalasit A, Scafidi M, Zuccarello B, Beghini M, Bertini L, Santus C, Valentini E, Benincasa A, Bertelli L (2010). The Hole-drilling Strain Gauge Method for the Measurement of Uniform or Non-uniform Residual Stresses, *AIAS Residual Stress Working Group*, TR01:2010, Florence, Italy.

- Oettel R (2000). The Determination of Uncertainties in Residual Stress Measurement (using the Hole Drilling Technique). Code of Practice 15, Issue 1, EU Project SMT4-CT97-2165.

Introduction of Strain Gauge Use

- Soete W, Vancrombrugge R (1950). An Industrial Method for the Determination of Residual Stresses. *Proc. SESA*, 8(1):17–28.

- Boiten RG, ten Cate W (1952). A Routine Method for the Measurement of Residual Stresses in Plates. *Applied Scientific Research*, A3(5):317–343.

- Riparbelli C (1950). A Method for the Determination of Initial Stresses, *Proc. of the SESA*, 8:173–196.

- Riparbelli C, Suppiger EW, Ward ER (1958). The Determination of Initial Stresses in Steel Plates. *Ship Structure Committee Report SSC-42*, U.S. Coast Guard, Washington DC.

- Rendler NJ, Vigness I (1966). Hole-drilling Strain-gage Method of Measuring Residual Stresses. *Experimental Mechanics*, 6(12):577–586.

- Beaney EM, Procter E (1974). A Critical Evaluation of the Center Hole Technique for the Measurement of Residual Stresses, *Strain*, 10(1):7–14.

Hole-Drilling Technique

- Flaman MT (1982). Brief Investigation of Induced Drilling Stresses in the Center-hole Method of Residual Stress Measurement. *Experimental Mechanics*, 22(1):26–30.

- Flaman MT, Herring JA (1986). Ultra-high-speed Center-hole Technique for Difficult Machining Materials. *Experimental Techniques*, 10(1):34–35.

- Beaney EM (1976). Accurate Measurement of Residual Stress on Any Steel Using the Center Hole Method. *Strain*, 12(3):99–106.

- Whitehead PS (2011). Practical Experiences in Hole Drilling Measurements of Residual Stresses. *Proc. of the SEM Annual Conference*, Society for Experimental Mechanics, 6(17):209–219. Indianapolis, June 7–10, 2010.

- Nau A, Scholtes B (2013). Evaluation of the High-speed Drilling Technique for the Incremental Hole-Drilling Method. *Experimental Mechanics*, 53(4):531–542.

- Steinzig M, Upshaw D, Rasty J (2014). Influence of Drilling Parameters on the Accuracy of Hole-drilling Residual Stress Measurements. *Experimental Mechanics*, 54(9):1537–1543.

Ring Core Method

- Meriam JL, DeGarmo EP, Jonassen F (1946). Method for the Measurement of Residual Welding Stresses. *Welding Journal*, 25(6):340s–343s.

- Milbradt KP (1951). Ring Method Determination of Residual Stresses. *Proc. of the SESA*, 9(1):63–74.

- Procter E, Beaney EM (1987). The Trepan or Ring Core Method, Center-Hole Method, Sach's Method, Blind Hole Methods, Deep Hole Technique. *Advances in Surface Treatments*, 4:165–198.

- Keil S (1992). Experimental Determination of Residual Stresses with the Ring-core Method and an On-line Measuring System. *Experimental Techniques*, 16(5):17–24.

- Ajovalasit A, Petrucci G, Zuccarello B (1996). Determination of Non-uniform Residual Stresses Using the Ring-core Method. *Journal of Engineering Materials and Technology*, 118(1):224–228.

- Hu Z, Xie H, Lu J, Zhu J, Wang H (2013). Determination of Nonuniform Residual Stresses Using the Ring-Core Method. *Measurement Science and Technology*, 24(8):085604.

Deep Hole Drilling

- Leeman ER (1968). A Technique for Determining the Complete State of Stress in Rock using a Single Borehole—Laboratory and Underground Measurements. *International Journal of Rock Mechanics and Mining Sciences and Geomechanics*, 5(1):31–38.

- Merrill RH (1967). Three-component Borehole Deformation Gage for Determining the Stress in Rock. U.S. Bureau of Mines, Report of Investigation 7015, 38pp.

- Beaney EM (1978). Measurement of Sub-surface Stress, *British Society for Strain Measurement Conference*, Bradford, UK.

- Leggatt RH, Smith DJ, Smith SD, Faure F (1996). Development and Experimental Validation of the Deep Hole Method for Residual Stress Measurement. *The Journal of Strain Analysis for Engineering Design*, 31(3):177–186.

- ASTM (2016). Standard Test Method for Determination of in Situ Stress in Rock Mass by Overcoring Method—Three Component Borehole Deformation Gauge. *Standard Test Method D4623-16*, American Society for Testing and Materials, West Conshohocken, PA.

- Amadei B, Stephansson O (1997). Methods of in Situ Stress Measurements. Chapter 2 in *Rock Stress and its Measurement*, Amadei B, Stephansson O, (Eds.), Springer, Dordrecht, Netherlands.

- Sjöberg J, Klasson H (2003). Stress Measurements in Deep Boreholes Using the Borre (SSPB) probe. *International Journal of Rock Mechanics and Mining Sciences*, 40(7-8):1205–1223.

Finite Element Calculations

- Beaney EM (1976). Accurate Measurement of Residual Stress on Any Steel Using the Center Hole Method. *Strain*, 12(3):99–106.

- Schajer GS (1981). Application of Finite Element Calculations to Residual Stress Measurements. *Journal of Engineering Materials and Technology*, 103(2):157–163.

- Nau A, von Mirbach D, Scholtes B (2013). Improved Calibration Coefficients for the Hole-drilling Method Considering the Influence of the Poisson Ratio. *Experimental Mechanics*, 53(8):1371–1381.

Incremental Stress Computation

- Bijak-Zochowski M (1978). A Semidestructive Method of Measuring Residual Stresses. *VDI-Berichte*, 313:469–476.

- Niku-Lari A, Lu J, Flavenot JF (1985). Measurement of Residual-stress Distribution by the Incremental Hole-drilling Method, *Experimental Mechanics*, 25(2):175–185.

- Schajer GS (1988). Measurement of Non-Uniform Residual Stresses Using the Hole-drilling Method. *Journal of Engineering Materials and Technology*, 110(4) Part I:338–343, Part II:344–349.

- Zuccarello B (1999). Optimal Calculation Steps for the Evaluation of Residual Stress by the Incremental Hole Drilling Method. *Experimental Mechanics*, 39(2):117–124.

- Schajer GS (2007). Hole-drilling Residual Stress Profiling with Automated Smoothing. *Journal of Engineering Materials and Technology*, 129(3):440–445.

- Schajer GS, Rickert TJ (2011). Incremental Computation Technique for Residual Stress Calculations Using the Integral Method. *Experimental Mechanics*, 51(7):1217–1222.

- Schajer GS, Winiarski B, Withers PJ (2013). Hole-drilling Residual Stress Measurement with Artifact Correction Using Full-Field DIC. *Experimental Mechanics*, 53(2):255–265.

Alternative Stress Calculation Methods

- Schajer GS (1991). Strain Data Averaging for the Hole-drilling Method. *Experimental Techniques*, 15(2):25–28.

- Kelsey RA (1956). Measuring Non-uniform Residual Stresses by the Hole Drilling Method. *Proc. of the SESA*, 14(1):181–194.

- Nickola WE (1986). Practical Subsurface Residual Stress Evaluation by the Hole Drilling Method, *Proc. of the SEM Spring Conference on Experimental Mechanics*, pp. 47–58, New Orleans, LA.

- Schajer GS, Yang L (1994). Residual-stress Measurement in Orthotropic Materials Using the Hole-drilling Method. *Experimental Mechanics*, 34(4):324–333.

Hole Eccentricity Compensation

- Sandifer JP, Bowie GE (1978). Residual Stress by Blind-hole Method with Off-center Hole. *Experimental Mechanics*, 18(5):173–179.

- Ajovalasit A (1979). Measurement of Residual Stresses by the Hole-drilling Method: Influence of Hole Eccentricity. *The Journal of Strain Analysis for Engineering Design*, 14(4):171–178.

- Beghini M, Bertini L, Mori LF (2010). Evaluating Non-uniform Residual Stress by the Hole Drilling Method with Concentric and Eccentric Holes. Part I and Part II, *Strain*, 46(4):324–336 and 46(4):337–346.

- Nau A, Scholtes B (2012). Experimental and Numerical Strategies to Consider Hole Eccentricity for Residual Stress Measurement with the Hole Drilling Method. *Materials Testing*, 54(5):296–303.

Material Plasticity Compensation

- Beghini M, Bertini L, Raffaelli P (1994). Numerical Analysis of Plasticity Effects in the Hole-drilling Residual Stress Measurement. *Journal of Testing and Evaluation*, 22(6):522–529.

- Beghini M, Bertini L, Raffaelli P (1995). An account of plasticity in the hole-drilling method of residual stress measurement. *The Journal of Strain Analysis for Engineering Design*, 30(3):227–233.

ESPI

- Nelson DV, McCrickerd JT (1986). Residual-stress Determination Through Combined Use of Holographic Interferometry and Blind-hole Drilling. *Experimental Mechanics*, 26(4):371–378.

- Steinzig M, Ponslet E (2003). Residual Stress Measurement Using the Hole Drilling Method and Laser Speckle Interferometry: Part I. *Experimental Techniques*, 27(3):43–46.

- Suterio R, Albertazzi A, Amaral FK (2006). Residual Stress Measurement Using Indentation and a Radial Electronic Speckle Pattern Interferometer—Recent Progress. *The Journal of Strain Analysis for Engineering Design*, 41(7):517–524.

Moiré

- McDonach A, McKelvie J, MacKenzie P, Walker CA (1983). Improved Moiré Interferometry and Applications in Fracture Mechanics, Residual Stress and Damaged Composites. *Experimental Techniques*, 7(6):20–24.

- Nicoletto G (1991). Moiré Interferometry Determination of Residual Stresses in the Presence of Gradients. *Experimental Mechanics*, 31(3):252–256.

- Wu Z, Lu J, Han B (1998). Study of Residual Stress Distribution by a Combined Method of Moiré Interferometry and Incremental Hole Drilling. *Journal of Applied Mechanics*, 65(4) Part I:837–843, Part II:844–850.

Digital Image Correlation

- McGinnis MJ, Pessiki S, Turker H (2005). Application of Three-dimensional Digital Image Correlation to the Core-drilling Method. *Experimental Mechanics*, 45(4):359–367.

- Nelson DV, Makino A, Schmidt T (2006). Residual Stress Determination Using Hole Drilling and 3D Image Correlation. *Experimental Mechanics*, 46(1):31–38.

- Schajer GS, Winiarski B, Withers PJ (2013). Hole-drilling Residual Stress Measurement With Artifact Correction Using Full-field DIC. *Experimental Mechanics*, 53(2):255–265.

CHAPTER 4

Strain Gauge Technique: Method Description

"The hole-drilling method can identify in-plane residual stresses near the measured surface of the workpiece material. The method gives localized measurements that indicate the residual stresses within the boundaries of the drilled hole."

ASTM E837 2013 "Standard Test Method for Determining Residual Stresses by the Hole-drilling Strain-gage Method."

The strain gauge hole-drilling method involves drilling a small hole at the geometric center of a specially designed strain gauge rosette bonded to the surface of the specimen. The residual stresses originally present within the hole are then calculated from the measured strain reliefs. The measurement procedure includes the following steps.

1. Choose the strain gauge rosette type.

2. Prepare the specimen surface for gauge installation.

3. Install the strain gauge rosette.

4. Connect the rosette to the strain measurement instrumentation.

5. Prepare the drilling machine, setting up, and aligning.

6. Drill a hole at the geometric center of the strain gauge rosette in a series of pre-determined depth increments and recording of the strain gauge readings after each depth increment. Measure the diameter of the drilled hole.

7. Present the gauge data (identification and orientation details, relaxed strain data, and hole diameter) in preparation for the calculation of residual stresses from a data reduction calculation.

The following sections give details of these steps.

4.1   STRAIN GAUGE ROSETTE SELECTION

Hole-drilling rosettes typically contain three radial strain gauges arranged in rectangular format (0°–90°–225° or 0°–45°–90°) to produce three in-plane relieved strain components for the computation of the three stress components σ_x, σ_y, and τ_{xy}. ASTM E837 describes the use of three different rosette types to suit a range of measurement needs. Figure 4.1 illustrates the three rosette geometries. The most widely used rosettes are Type A and Type B, each containing three radial gauge elements, large solder pads and center-line alignment markers. Type A, which follows the geometry described by Rendler and Vigness (1966), is a general-purpose design appropriate for most measurement needs. Type B has all three strain gauges placed on the same side of the hole location and is useful for making measurements adjacent to obstacles or boundaries.

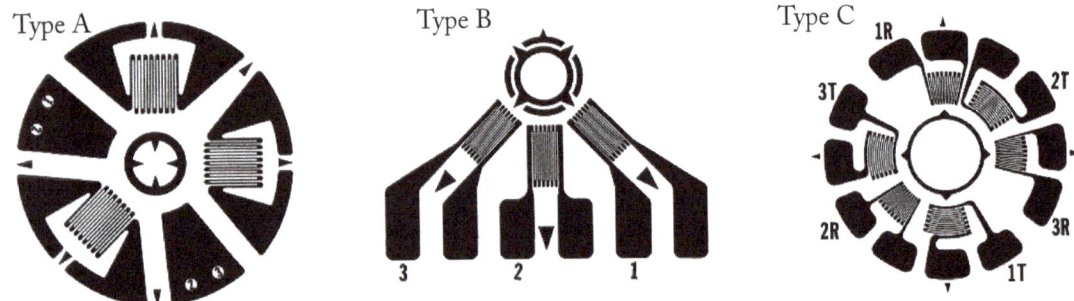

Figure 4.1: Standardized hole-drilling strain gauge rosettes (reproduced by permission of Micro-Measurements, a Vishay Precision Group brand).

Type A rosettes are available in three geometrically similar sizes to suit holes of nominal diameters 1/32", 1/16", and 1/8". Their layouts are identical, with all dimensions varying by a factor of 2 from one size to the next. In practice, the drilled hole diameters commonly used are larger than the nominal sizes, typically about 1 mm, 2 mm, and 4 mm, respectively. Typical hole depths equal half the drilled hole diameter. Type B rosettes are commercially available only in the 1/16" nominal size.

The Type C rosette is a specialized design suited to measurement of small residual stresses and to measurements on materials with low thermal conductivity such as plastics. The design comprises three radial strain gauges and three circumferential gauges, connected in three half-bridge circuits. This arrangement increases the effective strain sensitivity of the rosette and also provides compensation for thermal strains, both very useful features when measuring small strains. The thermal strain compensation also greatly stabilizes measurements on low-conductivity materials that do not provide adequate heat dissipation for strain gauges when connected within quarter-bridges. In general, this rosette pattern should be used only for these purposes because the half-bridges are costly and time consuming to assemble.

In general, strain gauges are manufactured either as "open" rosettes or with encapsulated grids. Various hole-drilling rosette patterns beyond the three types specified in E837 are commercially available from several different manufacturers and can also give satisfactory results providing that the associated calibration constants are used when computing the residual stresses from the measured strain data. Gauge manufacturers typically provide these calibration constants either explicitly or within available computer software.

It is usual for the selection of the strain gauge rosette to be based on the layout of the specimen, the size of the available target site and the depth to which residual stress data is required. The most commonly used rosette choice is the 1/16"-size. This rosette is drilled with 20 x 0.050 mm increments (as recommended in ASTM E837), which provides a good level of stress distribution detail and a final stress depth approaching 1 mm. If a greater level of stress distribution detail is required (especially close to the surface) then the 1/32"-size rosette (drilled with 20 x 0.025 mm increments to a final depth of 0.50 mm) may be more appropriate, provided that the hole can be drilled with the commensurate level of precision. Alternatively, if residual stress data is required to a greater depth, then the 1/8"-size rosette can provide data to depth 2 mm. Chapter 6 provides further details of gauge rosette layouts and sizes and examples of installations.

4.2 SPECIMEN PREPARATION

The quality of the gauge installation directly controls the quality of the strain readings measured. It is a requirement of the installation that, during the drilling process, all in-plane specimen displacements that occur over the gauge area are transferred to the gauge backing material by a continuous adhesive bond (or "glueline") that is as thin as practicably possible. To ensure that a sufficient level of adhesion is achieved between the specimen surface and strain gauge, the specimen must be prepared in an appropriate manner. Thus, the installation of the strain gauge rosette should be carried out in accordance with the gauge manufacturer's instructions. However, it is also to be noted that near-surface residual stresses within the specimen may form an important part of the data to be collected from the specimen and any surface treatments made to the specimen during preparation should be done in a manner that does not intrude on any required specimen features. An example installation is shown in Figure 4.2 where a strain gauge rosette has been bonded to a rough machined specimen. Show-through of the machining marks provides a "witness" at the gauge surface confirming that a thin glueline has been achieved here.

In cases where the specimen surface is highly irregular, corroded or otherwise unsuitable for direct installation of a strain gauge, some preliminary smoothing is required prior to final preparation for gauge installation. This can usually be done using machining or abrasive methods, but care should be taken to minimize such surface treatments because any pre-existing stresses close to the surface will be significantly redistributed by removal of surface material. In addition, it is important that the surface preparation method selected must not introduce spurious additional stresses into the newly formed surface, for example, by creation of excessive

Figure 4.2: Strain gauge rosette bonded to a rough-machined surface (photo courtesy of Stress-craft Ltd.).

temperatures or local plastic deformations. Surface preparation must be by a carefully controlled process using coolants wherever possible. As a general rule-of-thumb, the more rapid the material removal in a given surface preparation process, the greater the potential for spurious stress introduction. Ranked in order of least to most potential for spurious stress introduction (least to greatest material removal speed), the following processes may be applied to irregular specimens to produce surfaces for strain gauge installations:

- etching or super-polishing processes;

- coarse/medium grade silicon carbide paper (water cooled);

- coarse grade emery paper (dry); and

- flap wheel (low-speed and force).

More aggressive methods of material removal (for example, a high-speed abrasive disc) can impart residual stresses to depths in excess of 1 mm and should be used only in the last resort. The magnitudes and depths of spurious stresses introduced by various surface preparation methods will depend on the abrasion conditions, specimen material, thickness, etc. The affected material depth may be as little as 200 μm for a flap wheel or 50 μm for medium grade silicon carbide paper. In the case of etching or super-polishing processes, no additional stresses are introduced. However, these processes are not always readily available or easily controlled to produce the required results; some experimentation will be required to determine the characteristics of material removal rates using these methods.

Strain gauges can be satisfactorily bonded to almost any solid material if the material surface is prepared correctly. However, the key to surface preparation with regard to residual stress measurement is to develop a surface of sufficient roughness for strain gauge bonding without appreciably altering the state of the surface stresses. The surface for gauge bonding must have an appropriate surface roughness, be chemically clean and have a slight surface alkalinity. The steps of the surface preparation process recommended in Micro-Measurements Instruction Bulletin B-129-8 are:

- solvent degreasing: to remove oils and greases;

- surface abrading: to provide an adequate surface roughness;

- marking installation lines: to locate and orient the gauge accurately;

- surface conditioning: to remove all remaining dirt; and

- neutralizing: to return the surface to a slight alkalinity.

Bulletin B-129-8 suggests a surface finish (Ra) of between 1.6 and 3.2 μm and an alkalinity of between pH 7 and 7.5 for general stress analysis.

In practice, appropriate surface roughness is usually produced using mechanical abrasion. This can be achieved using fine silicon carbide paper (320–400 grade). Limiting the abrasion to a very small number of passes using new (sharp) paper for each gauge while applying light hand pressure prevents excessive heating and limits the introduction of spurious stresses to the specimen. Alternatively, chemical abrasion in the form of swab etching using an aggressive etchant may be used to produce surface roughness values approaching the required value while avoiding the introduction of spurious near-surface stresses. For example, acidic ferric chloride is very effective when working with many steel and nickel alloys. However, thorough neutralization is then required to remove all traces of acidic residues that may impede the subsequent gauge bonding process.

The marking of alignment lines is usually performed using a small, hard stylus, for example a fine, ball-point pen, followed by a solvent cleaning to remove any ink. If a scriber (or other sharp instrument) is used, then the resultant bur created at each side of the line should be removed using the abrasive paper used for surface preparation. Figure 4.3 shows typical alignment lines and a target strain gauge rosette.

4.3 GAUGE INSTALLATION

For most installations it is convenient to use cyanoacrylate adhesive to bond the strain gauge rosette to the specimen. Micro-Measurements Instruction Bulletin B-127-14 provides a highly detailed description of the use of this type of adhesive for bonding strain gauges. It is recommended that this document be studied carefully before installing a target gauge rosette and that some "practice" installations first be done to confirm that gauges can be aligned with the required

Figure 4.3: Alignment lines for a target strain gauge rosette (photo courtesy of Stresscraft Ltd.).

precision and that bonds of suitable quality are produced as a matter of routine. The requirement for a thoroughly planned approach and high level of cleanliness cannot be overemphasized. For completeness, the key steps in the procedure are given here.

a) Clean the surface of a glass slide (or gauge box lid) in preparation for placement of the gauge rosette.

b) Using tweezers, remove the rosette from its package and place it on the cleaned surface. Place a ~100 mm length of clear adhesive tape over the gauge (Figure 4.4).

c) Lift the tape and rosette from the surface and place them on the specimen target site. Adjust their position to align with the gauge center and specimen mark-out lines (Figure 4.4c).

d) Peel back one end of the tape and rosette from the specimen (Figure 4.4d).

e) Apply a small amount of catalyst to the rosette using a single brush stroke (Figure 4.4e) and allow it to dry.

f) Apply a single drop of adhesive to the tape fold adjacent to the rosette (Figure 4.4f).

g) Immediately after application of the adhesive, use a piece of protective film and roll in a single wipe the tape and rosette down onto the specimen (Figure 4.4g).

h) Maintain thumb pressure on the rosette until the bond has cured, typically ~ 1–2 min (Figure 4.4h).

i) Remove the tape by carefully peeling back (at acute angle to reduce lifting of the rosette; Figure 4.4i).

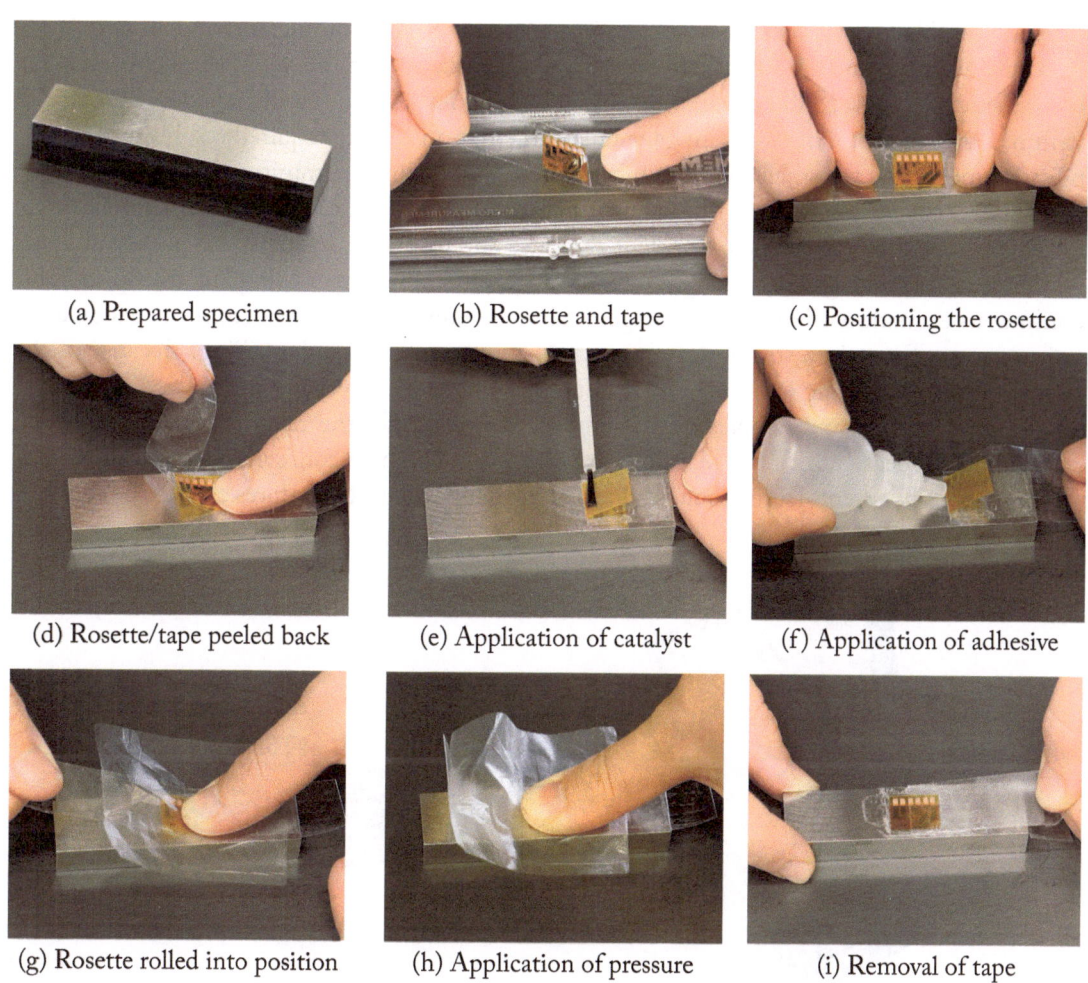

(a) Prepared specimen   (b) Rosette and tape   (c) Positioning the rosette

(d) Rosette/tape peeled back   (e) Application of catalyst   (f) Application of adhesive

(g) Rosette rolled into position   (h) Application of pressure   (i) Removal of tape

Figure 4.4: Target strain gauge installation on a test specimen (photos courtesy of Stresscraft Ltd.).

The installation can then be carefully wiped using a solvent to remove any excess adhesive and closely inspected.

Further details, including handling precautions for the materials involved in the application process, are described in Micro-Measurements Bulletin B-127-14. Some adaptation of the procedure may be required for less straightforward installations.

4.4   INSTRUMENTATION AND ELECTRICAL CONNECTIONS

Several portable and laboratory-based strain measuring and recording instruments are commercially available and are suitable for use with hole-drilling. ASTM E837 stipulates that the instrumentation for recording strains should have strain resolution, stability and repeatability all within $\pm\, 2\, \mu\varepsilon$. Most modern strain gauge instrumentation has the required resolution and stability for measuring the small strains in incremental hole-drilling. A minimum of three channels is required, one per gauge element or element pair. Continuous excitation of all channels is desirable because it avoids possible transient response and extended settling time as each channel is energised for strain gauge reading. For work on poor conductors or thin plate test specimens, continuous excitation offers the advantage of allowing any heating effect from the gauge excitation current to attain a steady state prior to commencement of hole drilling. Because of the time required for drilling each increment, there is no need for a very fast acquisition rate.

Figure 4.5 shows examples of instruments that can offer the required strain resolution, stability, and repeatability. They feature four channels that can be configured as quarter, half or full bridges. Instruments from other manufacturers that have similar features can also be suitable.

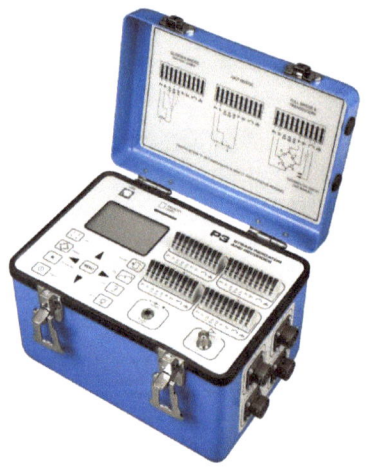   

(a) Vishay P3 Strain Indicator and Recorder   (b) Vishay D4 Data Acquisition Conditioner

Figure 4.5: Examples of strain measurement and recording instruments (reproduced by permission of Micro-Measurements, a Vishay Precision Group brand).

It is usual for electrical connections between the gauge and recording instrument to be made with the specimen fixed in the location where the hole drilling procedure is to be carried out. While not essential, this practice avoids placing any undue strain on soldered joints and gauge bonds from extraneous movements to and from the measurement location. Specimens can be glued to a work surface or base plate, or fixed using clamps. Care should be taken to ensure that any clamping loads do not introduce spurious loads at the target site.

For "open" gauge rosettes, electrical connections are first made by soldering single-strand conductors to the gauge elements (Figure 4.6a) and a remote solder terminal. Insulated, multi-stranded leadwires then link the solder terminal to the strain recorder. For encapsulated gauge rosettes, electrical connections are made by soldering insulated multi-strand leadwires to the gauge (Figure 4.6b) for direct connection to the strain recorder. Micro-Measurements Application Note TT-609 gives details of effective soldering techniques. It is recommended that the wires should be as short as practicable and that three-wire temperature-compensating circuits be used for each channel as described in Micro-Measurements Application Note TT-612. Shunt calibration of the circuits can be used to determine the impact of overall leadwire resistance on measured strains.

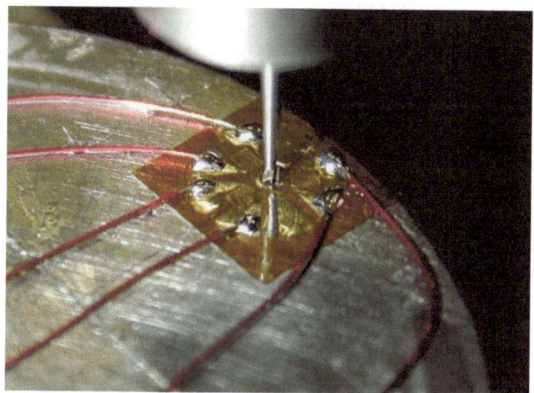

(a) Open gauge connections

(b) Encapsulated gauge connections

Figure 4.6: Examples of target strain gauge installations and leadwire connections (photos courtesy of Stresscraft Ltd.).

4.5 HOLE-DRILLING EQUIPMENT

The hole-drilling process is carried out using a specialized drilling guide or machine to align the drill axis with the strain gauge rosette, to provide the appropriate rotary motion and to control the drill depth. Readily available devices include the Vishay Micro-Measurements RS 200 and SINT MTS3000 machines pictured in Figure 4.7. Both machines incorporate optical systems for the alignment of the drill with the target center and for the measurement of the drilled hole diameter. They also incorporate dental-type air turbines for high-speed drilling. In addition, both machines use manual screws for the adjustment of the drill position over the target site.

The SINT machine works in conjunction with a personal computer to control the plunge (axial) drilling motion. A control panel and mouse/pointer enable selection of the drill increment parameters and provide control of the acquisition and recording of strain data. The Vishay driller is a "stand-alone" machine, where the drill feed is applied manually using a micrometer-type collar around the drill barrel. Most importantly, this machine can be configured so that the axis of the air-turbine is offset from the axis of the drill barrel. By rotation of the drill head within the barrel, drilling proceeds in an orbital manner. This is similar to trepanning (or circular milling) with a small eccentricity and can significantly enhance the quality of the hole forming process in comparison to plunge (axial) drilling. Figure 4.7 illustrates the main features of the two machines, while Figure 4.8 shows a schematic view of orbital drilling.

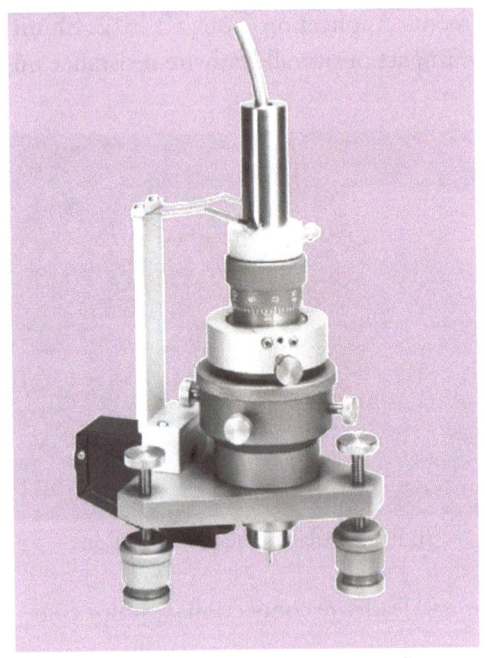

(a) Vishay Micro-Measurements RS-200
(reproduced by permission of
Micro-Measurements,
a Vishay Precision Group brand.)

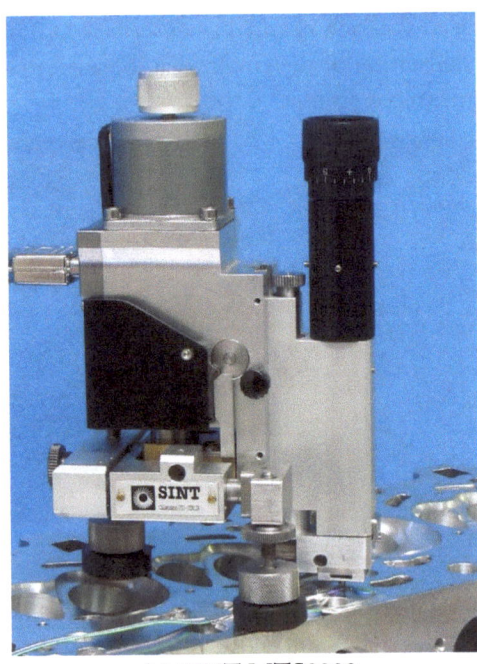

(b) SINT MTS3000
(photo courtesy of Sint Technology s.r.l.)

Figure 4.7: Hole-drilling machines.

For orbital drilling, the speed of rotation of the drill barrel about the hole center (green arrow) is, typically, less than 0.1% of that of the drill about its own axis (red arrow). The direction of orbit rotation is a matter for experimentation to suit the drill geometry and specimen material.

The recommended drill for most specimen materials is the inverted cone type in tungsten carbide. The inverted cone side profile provides clearance for the ejection of debris while avoiding rubbing at the sides of the drilled hole. The end of the drill should be flat or slightly concave.

Figure 4.8: Orbital drilling: a schematic view (photo courtesy of Stresscraft Ltd.).

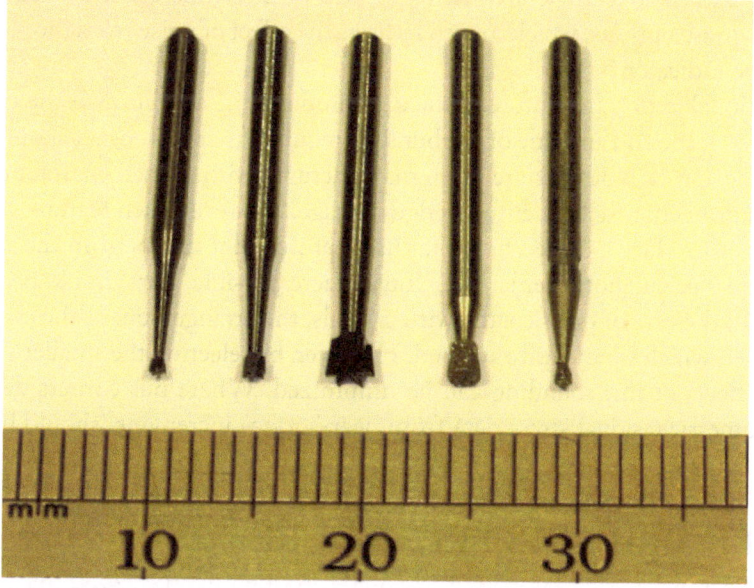

Figure 4.9: Inverted cone dental burs (photo courtesy of Stresscraft Ltd.).

Dental burs are a readily available source of drills suitable for hole drilling. Figure 4.9 shows a selection of inverted cone burs; the three drills to the left are of the tungsten carbide type while the two on the left are coated burs.

Coated burs can be useful for drilling into specimens whose hardness exceeds the machining capabilities of tungsten carbide. Cubic boron nitride coating can be used for drilling into hardened steels while diamond coatings are better suited to ceramic and glass-reinforced specimens. It is to be noted that coated drills are not well suited for plunge drilling because the stationary bur center does not cut well in an axial direction. An orbital drilling motion provides more satisfactory results.

Orbital drilling provides additional clearance at the side surfaces, but allowance must be made for the increase in diameter. For example, a 1.4-mm diameter drill and 0.3-mm radial orbit eccentricity may combine with bearing clearances and other tolerances to produce a final drilled hole diameter of ∼2.05 mm. It is recommended that test drillings (without a rosette) be made to establish the characteristics of the machine and specimen material to provide a satisfactory hole diameter within the range required by ASTM E837. These are defined as:

- 1/32" rosette: 0.93 mm to 1.06 mm diameter,

- 1/16" rosette: 1.88 mm to 2.12 mm diameter, and

- 1/8" rosette: 3.75 mm to 4.25 mm diameter.

Working with hole diameters at the upper end of these tolerances is desirable because it produces larger strain outputs, which in turn reduces the effect of noise from the gauge, leadwires and strain gauge indicator.

An additional important feature of the drill profile is the condition of the corner between the cutting edges at the end and side of the bur. It is required that this corner should be as "sharp" as possible. Tungsten carbide burs are often manufactured with a small chamfer or rounding at this corner as a protection against cutting edge damage. However, such features are undesirable here because they produce a drilled hole that has a profile that differs from the models used to generate the stress calculation coefficients. Thus, where possible, burs should be obtained with sharp corners. In the case of drills coated with crystals, the arrangement of the crystals produces a rounded corner, which is not easily avoided. However, by selecting the smaller sizes of coating medium, the effects of this rounding can be minimized. Where bur corners depart from the ideal sharp geometry, residual stress data from near-surface increments should be treated with caution.

4.6   HOLE-DRILLING PROCEDURE

In practice, hole drilling proceeds in the steps below. While the description is most applicable to manual control using an RS-200 drilling machine, most details also apply to other types of machines.

- Set the strain recording instrument parameters to suit the bridge connections, strain gauge factors, etc.

- Check the method for holding the specimen to ensure that no movement can take place during the drilling process. If clamps are used to secure the specimen, these should be adjusted while monitoring the strain gauge outputs to confirm that no significant loads are applied at the gauge site.

- Place the drilling machine on the work surface in a position approximately centered over the target gauge. Fix the three supporting feet in position (usually by cementing) to hold the drilling machine firmly in place. Then make adjustments to the vertical height of each of the three mounting screws to make the drilling axis perpendicular to the specimen surface at the target center. This last adjustment can be checked using small squares or devices inserted into the drill barrel and brought close to gauge surface to view any asymmetry. The vertical height screws should then be locked.

- Insert the optical head into the drill barrel and focus it onto the gauge. Adjust the x-y position screws so that the cross-hairs of the optical head coincide with the center of the target gauge. If possible, rotate the eyepiece containing the cross-hairs within its mounting to check that the cross-hairs are accurately centered. Then remove the optical head from the driller.

- Inspect the selected drill using a magnifying eyepiece. If needed, also measure its diameter using the optical head. Defective drills, e.g., those that are chipped, rounded, or with excessive chamfers, should be rejected. Figure 4.10 shows examples of drills during inspection. When ready, insert the inspected drill into the drill motor collet and tighten it to the extent recommended by the manufacturer.

- Taking care not to touch the strain gauge rosette, assemble the drill motor into the drill barrel and lock it into position. Switch on the air supply and run the turbine up to speed. Observe the gauge strain outputs to ensure that there are no significant thermal effects from the exhaust air temperature differences.

- Slowly turn the micrometer feed to lower the rotating drill toward the gauge. A change in the sound of the motor indicates when the drill touches and then drills through, the gauge backing material. When the drill makes contact with the specimen, the drill depth datum is established. At this point, the feed is stopped, the micrometer reading recorded and the air supply switched off. Oblique viewing of the drill/gauge/specimen during this process (using a magnifying eyepiece) can assist in the determination of the drill depth datum (Figure 4.11).

- Balance the gauge bridges so that all three strain elements show a reading of zero.

(a) Drill with corner chamfers

(b) Drill with corner chamfers

(c) Drill with "sharp" corners

(d) Drill with "sharp" corners

Figure 4.10: Inspection of new drills (prior to drilling process) (photos courtesy of Stresscraft Ltd.).

- Drill the hole in a series of depth increments. The increments are defined in ASTM E837 as 20 x 0.05 mm for 1/16"-size gauges, 20 x 0.025 mm for 1/32"-size gauges, or 20 x 0.10 mm for 1/8"-size gauges. For each increment, start the air turbine and advance the micrometer feed in a number of sub-increments while rotating the drill assembly within the drill barrel to produce the required orbital machining motion. Allow the drilling process to "run-out" (complete a full orbit) at each sub-increment to ensure that the machining is complete. On reaching the end of the depth increment, switch off the air supply to the air turbine and observe the strain outputs. Continue to observe the settling of the output values to note the dissipation of any transient thermal response. Record the strains for each

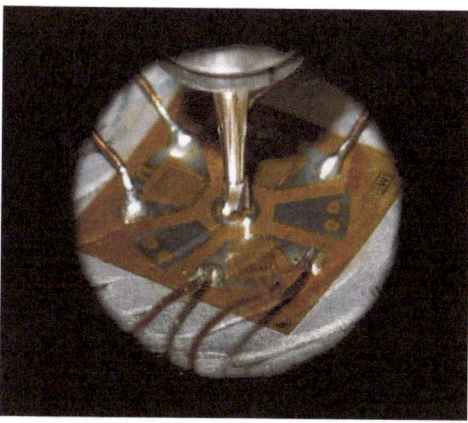

Figure 4.11: Oblique view of drilling process (photo courtesy of Stresscraft Ltd.).

increment when steady values have been established. Repeat this process until the final hole depth has been attained.

- On completion of the final drilling increment, use the micrometer feed screw to lift the drill motor assembly and remove the drill from the hole. The motor assembly can then be removed from the drill barrel.

- Insert the optical head into the drill barrel and make a preliminary inspection of the target area and gauge to confirm the concentricity of the drilled hole and gauge target center. Remove the optical head.

- Unsolder the leadwires from the gauge and peel the gauge from the specimen using a sharp blade. Where possible, avoid making contact between the blade and specimen surface around the hole.

- Reinsert the optical head. Measure the diameter of the hole using the optical graticule in both x- and y-directions. Comparison of the two measurements can help in the detection of any drilling problems; use the average value for subsequent calculations. In addition, inspect the edge of the hole to confirm that no upward burs are present. If present, such burs indicate the onset of plastic deformations at the start of the drilling process. Figure 4.12 shows examples of holes following drilling and removal of the gauge. The depth of the hole can be measured as a final check, if required.

- Finally, remove the used drill from the air turbine collet and inspect it using a magnifying eyepiece or the optical head. Figure 4.13 shows a drill bit with typical wear following drilling into a "tough" specimen. The likely impact of the observed wear or damage to the cutting edges on the recorded strains or hole profile are a matter for the practitioner's judgment. Except when used for drilling very soft materials, drills should be discarded after completion of one hole and a new drill used for the next hole.

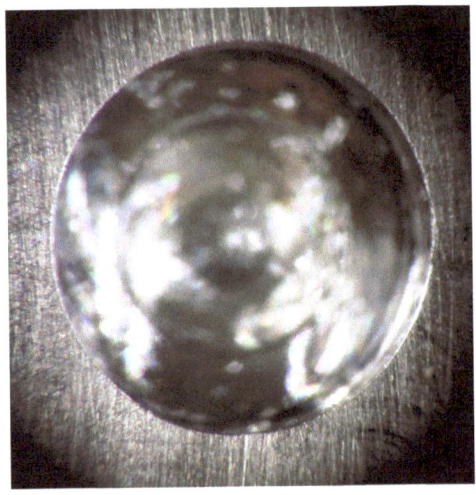

(a) Hole with "clean" edge

(b) Hole with blurred edge

Figure 4.12: Inspection of drilled holes (photos courtesy of Stresscraft Ltd.).

Figure 4.13: Inspection of a worn drill after completing the drilling process (photo courtesy of Stresscraft Ltd.).

4.7   GAUGE DATA

It is recommended that gauge data are recorded on a standard form. This form should contain the following information:

- the date and time of the test;

- the specimen identification details (material, surface treatment, etc.);

- the material properties (Young's modulus and Poisson's ratio);

- the gauge identification number;

- the drilling machine identification number (if appropriate);

- the strain measurement/recording instrument identification number;

- the strain gauge type, lot number and gauge factor;

- the array of hole depths and relaxed strains at those depths; and

- the final hole diameter (and depth, if required).

After completion of all measurements and associated documentation, the residual stress values corresponding to the measured strains can be calculated as described in Chapter 5.

The procedures described in this chapter are suitable for the majority of routine hole-drilling residual stress measurements. However, it often happens that some feature of the specimen or measurement objective requires a procedural variation. Such variations can be accommodated within the basic hole-drilling technique providing that care is taken to preserve the integrity of the overall measurement process. Chapter 6 describes some example applications that require procedural adaptations to meet their measurement objectives.

4.8   FURTHER READING

- ASTM-E837 (2013). Determining Residual Stresses by the Hole-drilling Strain-gage Method. *ASTM Standard Test Method E837-13a*. American Society for Testing and Materials, West Conshohocken, PA.

- Rendler NJ, Vigness I (1966). Hole-drilling Strain-gage Method of Measuring Residual Stresses. *Experimental Mechanics*, 6(12):577–586.

- Vishay Measurements Group, Inc. (1993). Measurement of Residual Stresses by the Hole-drilling Strain-gage Method. *Tech Note TN-503-6*. Vishay Measurements Group, Inc., Raleigh, NC. 16pp.

• Grant PV, Lord JD, Whitehead PS (2002). The Measurement of Residual Stresses by the Incremental Hole Drilling Technique. *Measurement Good Practice Guide*, No. 53, National Physical Laboratory, Teddington, UK.

• Vishay Precision Group, Inc. (2011). Surface Preparation for Strain Gauge Bonding. *Instruction Bulletin B-129-8*, December 19, 2011.

• Vishay Precision Group, Inc. (2011). Strain Gage Installations with M-Bond 200 Adhesive. *Instruction Bulletin B-127-14*, October 14, 2011.

• Vishay Precision Group, Inc. (2010). Strain Gage Soldering Techniques. *Application Note TT-609*, November 13, 2010.

• Vishay Precision Group, Inc. (2010). The Three-wire Quarter-bridge Circuit. *Application Note TT-612*, November 14, 2010.

CHAPTER 5

Stress Computations

"The residual stresses originally existing at the hole location are evaluated from the strains relieved by hole-drilling using mathematical relations based on linear elasticity theory."

ASTM E837 (2013) "Standard Test Method for Determining Residual Stresses by the Hole-Drilling Strain-Gage Method."

5.1   INTRODUCTION

After the hole has been drilled and the associated surface deformations measured, the remaining major procedural step is to compute the indicated residual stresses. Even in the simplest case where a single average local residual stress value is desired, the calculation requires some sophistication. This need occurs because hole-drilling involves partial strain relief only, so the specific fraction of the strain relief needs to be known. This is achieved using theoretical or experimental calibrations. For more detailed measurements, where the within-depth profile of the residual stresses is the objective, a much more complex calculation procedure becomes necessary.

This chapter describes some practical methods for computing hole-drilling residual stresses from measured surface deformation data. It starts from the basic case where the residual stresses are uniform through the depth of the specimen material, and goes on to the much more challenging case of evaluation of within-depth residual stress profiles. The given details refer to computation procedures suitable for use with the strain data measured using strain gauges. Chapter 7 describes analogous procedures specialized for use with full-field deformation data from optical methods such as ESPI, Moiré and DIC.

5.2   UNIFORM RESIDUAL STRESSES

The Hole-Drilling Method measures the residual stresses that originally existed within the material cut from the drilled hole. This material lies next to the specimen surface, where plane-stress conditions occur. Consequently, the out-of-plane stresses are very small and may be considered negligible. Hole drilling measurements therefore focus on the three in-plane stresses σ_x, σ_y and τ_{xy}.

"Uniform" residual stresses are said to occur when the stresses that originally existed within the material cut from the hole were of similar size throughout the depth and diameter of the hole. In this case, the set of three in-plane stress values fully describe the local residual stress state. Evaluation of these three stresses requires three independent measurements, for example, provided by a three-element strain gauge rosette similar to that shown in Figure 5.1.

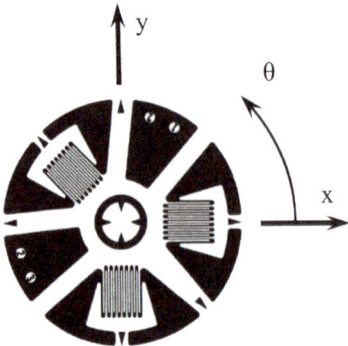

Figure 5.1: Typical hole-drilling strain gauge rosette (reproduced by permission of Micro-Measurements, a Vishay Precision Group brand).

Commercial hole-drilling rosettes are universally of the 45° rectangular style because this arrangement simplifies the associated stress evaluations. The greatest measurement accuracy of the in-plane strain state is achieved when the gauges 1 and 3 in Figure 5.1 are aligned with the principal strain directions (same as principal stress directions in an isotropic material). Thus, when installing a 45° rosette, it is preferable, although not essential, to align it with the principal stress directions, if they are known.

It is common to label the two rectangular strain gauges of a 45° rectangular rosette as the x and y axes, as shown in Figure 5.1. In this arrangement, gauge 1 aligns with the x axis. For an isotropic material whose boundaries are far from the hole location, the strain $\varepsilon(\theta)$ measured by a strain gauge aligned at angle θ counter-clockwise from the x direction follows a trigonometric relationship:

$$\epsilon(\theta) = \left(\sigma_x + \sigma_y\right) A + \left(\sigma_x - \sigma_y\right) B \cos 2\theta + \tau_{xy} 2B \sin 2\theta, \tag{5.1}$$

where A and B are calibration constants whose numerical values depend on the diameter and depth of the drilled hole and the elastic properties of the specimen material. In practice, these calibration constants are determined either experimentally or numerically.

Equation (5.1) also applies to the strains observed in the Ring-Core Method. For the case where the ring-core is cut to a great depth in a uniformly stressed material, the cutting of the ring-core fully relieves the residual stresses and the calibration constants A and B can be determined by application of Hooke's Law to be

$$A = -\frac{1 - \nu}{4E} \qquad B = -\frac{1 + \nu}{4E}. \tag{5.2}$$

The leading minus signs occur because the strain measurements are made during material unloading. Equations (5.2) show that the constants A and B are material dependent, with an inverse proportionality to Young's modulus, E. The same inverse proportionality with Young's modulus occurs with hole-drilling. However, the Hole-Drilling Method differs importantly from the Ring-Core Method in that the strain measurement locations are remote from the associated residual stress locations. This aspect of the method causes the constants A and B to have greatly different numerical values and for the relationships with Poisson's ratio to deviate from those in Equations (5.2). The classical Kirsch analytical solution for the stresses and strains around a hole in a stressed material indicates a $(1 + v)$ influence on the hole-drilling constant A and a much weaker influence of Poisson's ratio on the hole-drilling constant B. The latter effect is so small that it may reasonably be ignored for Poisson's ratios within the narrow range found in engineering materials. With these observations in mind, it is convenient to express the hole-drilling calibration constants in dimensionless form. This generalizes their application beyond the specific material for which they were originally derived to any linear elastic material obeying Hooke's Law.

$$\bar{a} = \frac{2AE}{1 + v} \qquad \bar{b} = 2BE \tag{5.3}$$

where the superscript bar notation is an historical remnant from previous practice where the calibration constants were sometimes evaluated by considering the strain at the center point of each strain gauge. This was only a modestly accurate approach because the strain field within the strain gauge boundaries is highly nonlinear, causing the center strain to deviate significantly from the overall average strain. The superscript bar notation was introduced to indicate that these calibration constants are determined from consideration of the whole strain gauge area.

Average strain can be computed from the double integral of the pointwise strain over the length and width of the strain gauge. However, it happens that the integral of strain is displacement. Thus, the average strain can be more compactly found from the single integral of relative end displacements in the direction of the gauge conductors across the strain gauge width. This is a particularly convenient approach when working with the results of finite-element calculations because these results naturally appear in terms of displacements. The redundant further steps of differentiating those displacements to produce strains and then integrating the results back to displacements can entirely be avoided.

In terms of the dimensionless calibration constants $\bar{a}$ and $\bar{b}$, Equation (5.1) becomes

$$\varepsilon(\theta) = P\bar{a}(1 + v)/E + Q\bar{b}\cos 2\theta/E + T\bar{b}\sin 2\theta/E, \tag{5.4}$$

where, for the convenience of the subsequent computation procedures, the residual stresses have been expressed in combined form as

$$P = \left(\sigma_x + \sigma_y\right)/2, \quad Q = \left(\sigma_x - \sigma_y\right)/2, \quad T = \tau_{xy}. \tag{5.5}$$

Stress P is the isotropic (direction independent) component of the in-plane stresses. It is the 2-D version of a hydrostatic (pressure type) stress. Stresses Q and T are deviatoric (shear type) stresses, respectively, the shear stresses in the 45° and the x-y directions.

The strain gauges in a standard hole-drilling rosette are aligned at angles $\theta = 0°, 135°$ and 270°. Substituting these values into Equation (5.4) gives

$$\varepsilon_1 = P\bar{a}(1+v)/E + Q\bar{b}/E$$
$$\varepsilon_2 = P\bar{a}(1+v)/E - T\bar{b}/E \qquad (5.6)$$
$$\varepsilon_3 = P\bar{a}(1+v)/E - Q\bar{b}/E.$$

When inverted to allow evaluation of the residual stresses from the measured strains, Equations (5.6) become

$$P = \frac{E}{\bar{a}(1+v)}p \quad Q = \frac{E}{\bar{b}}q \quad T = \frac{E}{\bar{b}}t, \qquad (5.7)$$

where the combined strains p, q, and t are

$$p = (\varepsilon_1 + \varepsilon_3)/2 \quad q = (\varepsilon_1 - \varepsilon_3)/2 \quad t = (\varepsilon_1 + \varepsilon_3 - 2\varepsilon_2)/2. \qquad (5.8)$$

These strains are directly analogous to the corresponding stresses P, Q and T. Strain p is the isotropic (direction independent) component of the in-plane strain. Strains q and t are deviatoric (shear type) strains, respectively, the shear strains in the 45° and the x-y directions.

Finally, the Cartesian stresses and the principal stresses and directions can be determined from P, Q, and T using

$$\sigma_x = (P+Q) \quad \sigma_y = (P-Q) \quad \tau_{xy} = T \qquad (5.9)$$

$$\sigma_{max}, \sigma_{min} = P \pm \sqrt{Q^2 + T^2} \qquad (5.10)$$

and where the direction of the more tensile (or less compressive) principal stress is

$$\beta = 0.5^* \arctan(-T/-Q), \qquad (5.11)$$

where β is measured clockwise from the direction of gauge 1 (x-direction) in Figure 5.1. The second principal stress (more compressive or less tensile) is in the perpendicular direction, that is, at angle β measured clockwise from gauge 3 (y-direction) in Figure 5.1 (either the $+y$ or $-y$ directions can be used). Use of the two-argument arctangent function places β in the correct quadrant, for example ATAN2(-Q,-T) in Excel, atan2(-Q,-T) in C and Matlab and ATAN2(-T,-Q) in Fortran. Note that the various computer languages use the arguments in different orders. The seemingly redundant minus signs should be retained because the two-argument arctangent function separately uses the signs of the numerator and denominator to place the resulting angle in the correct quadrant. Table 5.1 illustrates the angular placement resulting from Equation (5.11).

Table 5.1: Placement of the principal angle β

	Q > 0	Q = 0	Q < 0
T < 0	45° < β < 90°	45°	0° < β < 45°
T = 0	90°	undefined	0°
T > 0	-90° < β < -45°	-45°	-45° < β < 0°

5.3 CALIBRATION CONSTANTS

In the early years of hole drilling, the numerical values of the calibration constants $\bar{a}$ and $\bar{b}$ were evaluated through experimental measurements on specimens with known stresses. The values of the constants vary with hole depth and diameter, also the size and geometrical arrangement of the strain gauge rosette used for the measurement. Rendler and Vigness (1966) were the first to note that the strain response depends on the ratio of the hole dimensions to the strain rosette size, for example, as indicated by the mean diameter of the strain gauge circle. This observation introduced the concept of normalization relative to rosette size so that calibration constants could be used with different sizes of strain gauge rosettes, providing that the rosettes are geometrically similar. This is the case with the "Type A" rosettes specified in ASTM E837, which are available in 1/32", 1/16", and 1/8" nominal sizes. It is now standard practice to quote calibration constants in terms of hole depth and diameter as fractions of the mean diameter of the strain gauge rosette.

In modern practice, the calibration constants are evaluated using finite element computations. The results are more consistent and reliable than experimentally derived values. In addition, the computational approach can evaluate calibration constants for stress states that are difficult or impossible to create experimentally. The fundamental analytical procedure for determining hole-drilling calibration constants uses an approach independently introduced by Boiten and ten Cate (1952) and Schajer (1981). It is essentially an application of Bueckner's Principle, which was stated in 1958 for crack stress problems and which states "Any [elastic] crack or notch problem can be reduced to one where the external load appears in the form of tractions distributed over the faces of the crack." Figure 5.2 illustrates the procedure for hole-drilling computations.

Figure 5.2a schematically shows a cross-sectional view of the stress state at the measurement location before the hole is drilled. The diagram seems to show the state after the hole has been drilled. However, the original undrilled stress state has been restored in the surrounding material by applying the originally existing stresses at the hole boundary.

Figure 5.2c represents the stress state at the measurement location after the hole has been drilled. As is the case in practice, the hole boundary is stress free and only the far-field is stressed. Using the principle of superposition, the stress state in Figure 5.2b can be added to the "before drilling" stress state in Figure 5.2a to create the "after drilling" stress state Figure 5.2c. Thus, a

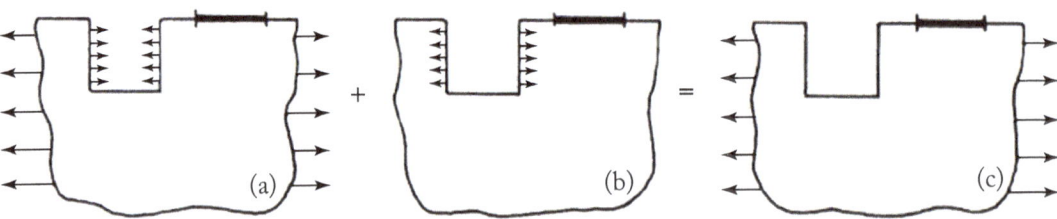

Figure 5.2: Superposition of stress states to evaluate effect of hole-drilling stress relief (from Schajer (1981)).

finite element analysis of the stress state in Figure 5.2b will model the local deformation caused by the hole drilling process and will therefore directly provide the calibration constants $\bar{a}$ and $\bar{b}$.

Apart from its computational convenience, the superposition in Figure 5.2 importantly demonstrates that the Hole-Drilling Method provides a localized measurement. It measures the residual stresses that originally existed at the hole boundary, even though the actual measurements are made over a much more extended area in the surrounding material. Thus, while it is necessary for the original residual stresses to have been uniform within the diameter of the hole, it is not required that they were also uniform outside the hole, for example, over the strain gauge area in Figure 5.1. This observation allows localized residual stress measurements to be made, for example in a weld, where the drilled hole must be kept within the weld metal but the strain gauges may be allowed to overlap into the adjacent material.

Hole-drilling calibration constant $\bar{a}$ relates the hydrostatic stress P to the hydrostatic strain p. As seen from Equation (5.4), there is no variation with angle θ, so an axisymmetric 2D analysis is sufficient to provide a full solution for this case. Figure 5.3 shows an example finite element mesh used to model the cross-section shown in Figure 5.2b. A pressure load is applied along the hole boundary to represent the residual stresses. The far-field boundary has no applied stresses, but it is important that it realistically models the far-field boundary conditions. This can be done by modeling a cross-section extending to a very large radius. Recent work by Baldi has shown that good results can be achieved while using a much smaller outside radius by adding a region of high stiffness at the distant boundary to represent the effect of an "infinite" far-field.

It happens that a similar 2-D approach can also be used to evaluate hole-drilling calibration constant $\bar{b}$, which describes the response to the shear stresses Q and T. In this case, Equation (5.4) indicates a $\cos 2\theta$ and $\sin 2\theta$ variation with angle. Since these relationships are explicitly known, there is no need to discretize them, for example, through a 3-D finite element analysis. Instead, 2-D "harmonic" elements that have the $\cos 2\theta$ and $\sin 2\theta$ variations built into them can be used. This approach, using a separate 2-D finite element analysis for each of the calibration constants $\bar{a}$ and $\bar{b}$, gives substantial computational advantage over a 3-D analysis because it uses far fewer elements, so allowing more detailed discretization and hence greater

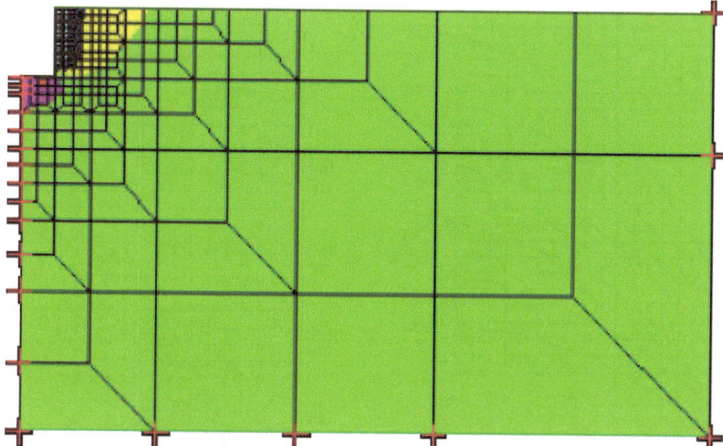

Figure 5.3: Example 2-D finite element mesh for computing hole-drilling calibration constants.

accuracy. Contrary to occasional statements, the method does not involve any approximation for the most common case of a linear elastic isotropic material, where the specimen is large compared with the hole size. Only when the material is not linear elastic, for example, due to plasticity, or it is not isotropic, or the specimen boundaries are near the hole, does a full 3-D analysis become necessary.

The various strain gauge rosettes produced by different manufacturers are generally similar, but vary in some details. Thus, their calibration constants are also similar, but are not identical. For the most reliable stress computations, it is desirable to use the calibration constants corresponding to the particular strain gauge rosette used. These data are typically provided by the manufacturer in graphical or numerical form or contained within computer software. Figure 5.4 shows an example graphical presentation of the calibration constants $\bar{a}$ and $\bar{b}$ for an ASTM E837 Type A rosette. Table 3 in ASTM Standard Test Method E837-13 presents these data in numerical form, also the data for rosette Types B and C. The presented values are for the case where the specimen is much larger than the strain gauge rosette, with thickness greater than the mean rosette diameter and distance to the nearest boundary at least 1.5 times the mean rosette diameter.

The curves in Figure 5.4 show that the calibration constants vary with hole diameter and with hole depth. The results are presented with these hole dimensions normalized relative to the mean diameter of the strain gauge rosette. With this normalization, the strain vs. hole depth responses become approximately proportional to the square of the hole diameter. This feature simplifies graphical presentation and subsequent interpolations between presented diameter values.

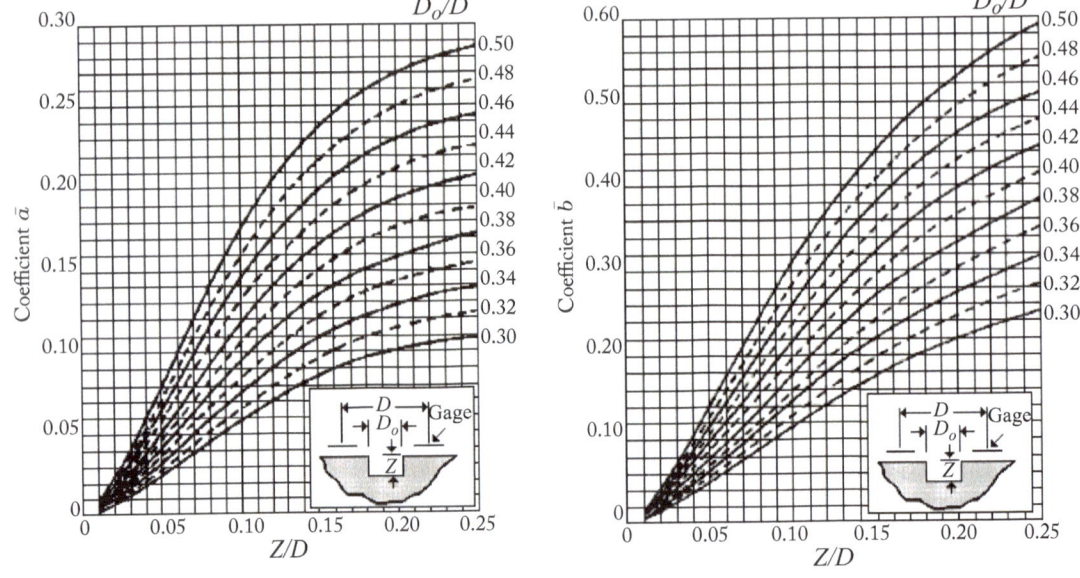

Figure 5.4: Hole-drilling calibration constants $\bar{a}$ and $\bar{b}$ for an E837 Type A rosette (reproduced by permission of Micro-Measurements, a Vishay Precision Group brand).

5.4   STRESS AVERAGING

Conceptually, a "uniform" residual stress measurement can be completed by measuring the strain changes caused by drilling a hole directly to its final target depth and then using Equations (5.7)–(5.11) with the calibration constants from Figure 5.4 or other source to determine the associated residual stresses. Three strains are measured from the strain gauge rosette and are used to evaluate the three in-plane components of stress. This is the original form of the Hole-Drilling Method. However, this approach uses a minimum of data and does not take advantage of further data that could be used.

The choice of "target" hole depth is left open to the user. Conceptually, any target depth is acceptable providing the corresponding calibration constants are used for the residual stress calculation. The larger hole depths close to the right sides of the graphs in Figure 5.4 are attractive because they give larger measured strains. However, the smaller relieved strains at smaller hole depths can also be useful as additional information. The responses at both smaller and larger hole depths can be combined by choosing to cut the hole in a series of small depth increments and measuring the relieved strains after each one. This provides a history of the evolution of the relieved strains with hole depth such as shown in Figure 5.5. The residual stress calculation takes the average of the stresses computed at each step. More reliable results are expected from the strains measured at the larger hole depths, so it is desirable to bias the calculations toward the greater depth measurements. This can conveniently be done by weighting the averaging using the calibration constant at each hole depth. The stress calculation in Equation (5.7) is then replaced by

$$P = \frac{E}{(1+v)} \frac{\sum(\bar{a} \cdot p)}{\sum(\bar{a})^2} \quad Q = E\frac{\sum(\bar{b} \cdot q)}{\sum(\bar{b})^2} \quad T = E\frac{\sum(\bar{b} \cdot t)}{\sum(\bar{b})^2}. \tag{5.12}$$

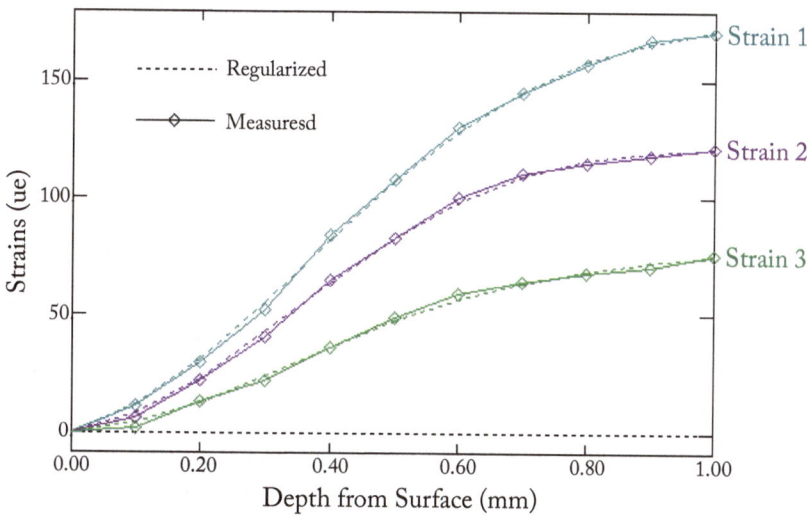

Figure 5.5: Evolution of relieved strains measured with incremental hole drilling.

ASTM E837-13 specifies the use of 20 hole depth increments of 0.05 mm to a final depth of 1 mm when using a 2 mm diameter hole in a 1/16" nominal Type A rosette. Deeper holes would provide larger strains, but the Standard limits the hole depths to half a hole diameter to reduce stress concentration effects so as to allow evaluations of residual stresses nearer the yield stress.

Another advantage of making strain measurements over a series of hole depth increments is that it reveals the evolution of the relieved strains. This evolution should be smooth, as in Figure 5.5. Thus, if it should happen that some fault occurs during the measurement process, it would likely create a visible anomaly in the strain vs. depth data. This eventuality could be handled by the removal of isolated faulty data or by repetition of the entire measurement should the fault be extensive.

Equations (5.12) are designed to be applied to the evaluation of uniform residual stresses. However, they are sometimes applied to non-uniform stresses so as to produce a comparative measure of their general "size." The results are called the Equivalent Uniform Stresses (EUS), and correspond to the uniform stress state that would produce the same response with Equations (5.12). The EUS is a weighted average of the non-uniform residual stresses within the hole depth, with the result most influenced by the stresses near the surface.

5.5   NON-UNIFORM RESIDUAL STRESSES

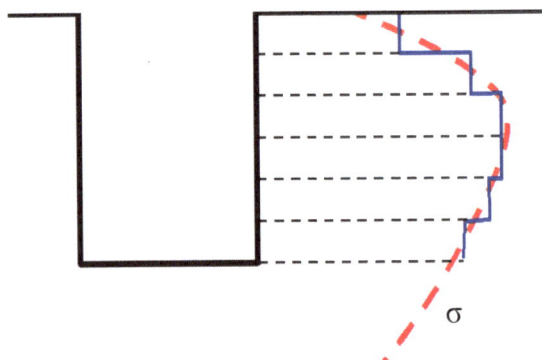

Figure 5.6: Schematic diagram of the profile of residual stress that varies with depth from the specimen surface (from Schajer and Rickert (2011)).

In general, residual stresses in thick specimens are not uniform through the depth of a material. The dashed red line in Figure 5.6 schematically shows an example where the residual stress varies with depth from the specimen surface. The relationship between these residual stresses and the resulting surface strains measured during hole-drilling depend on two factors:

1. **Stress Effect**. All the stresses at the various depths within the overall hole depth contribute to the measured surface strains. The greatest contributions come from the stresses near the measured surface. The interior stresses contribute progressively less with distance from the measured surface. Beyond about one hole radius the contribution becomes so small that it is no longer useful when seeking to identify residual stresses from surface strain measurements. This effect limits the maximum stress depth range of the Hole-Drilling Method. It is a physical limitation of the measurement technique, not a mathematical artifact of the residual stress evaluation method. It occurs because the residual stresses are self-equilibrating and so have a zero force resultant. Thus, St. Venant's principle indicates that their far-field effect diminishes to zero.

2. **Geometric Effect**. The creation of the hole progressively reduces the stiffness of the surrounding material. Thus, the surface strains produced by a stress at a given depth generally increase as the hole becomes deeper. Small reversals can occur at large hole depths through complex geometric interactions. The geometric effect is clearly demonstrated in hole-drilling measurements on plated materials that have a highly stressed surface plating

attached on an approximately stress-free substrate. Surface strains continue to increase after the hole has penetrated the stressed plating material and proceeds into the (relatively) stress-free substrate. The same effect occurs in homogeneous materials.

Equation (5.13) illustrates the mathematical relationship between the measured surface strains and the interior residual stresses. The example equation refers to the case relating the "p" (isotropic) strains measured at hole depth h and the corresponding "P" stresses that exist at the various depths H within the hole depth h. Similar equations relate the deviatoric strain and stress quantities q, Q and t, T:

$$p(h) = \frac{1+v}{E} \int_0^h \widehat{A}(H,h)P(H)\,dH \qquad 0 \le H \le h. \tag{5.13}$$

Equation (5.13) is classified as a Volterra equation of the first kind. Its integral format reflects the action of the stress effect, where all stresses within the hole depth combine to contribute to the surface strains. The kernel (strain response) function $\widehat{A}(H,h)$ represents the geometric effect. It describes the surface strain response per unit depth caused by a unit stress at depth H, when the hole depth is h.

Equation (5.13) is categorized as an "inverse problem" because the quantity to be determined appears on the right side, within the scope of the integral. This feature significantly complicates the required solution procedure. In contrast, a conventional "forward problem" can be solved simply because the quantity to be determined appears individually on the left, with the solution appearing explicitly on the right.

Equation (5.13) is expressed in terms of continuous functions. However, the strain response, $p(h)$, is measured experimentally only at a finite number of discrete hole depths h_i, $i = 1, 2, \ldots, n$. In this case, a practical stress solution can be achieved by recasting Equation (5.13) in discrete form:

$$p_i = \frac{1+v}{E} \sum_{j=1}^{j=i} \bar{a}_{ij}\, P_j \qquad 1 \le j \le i \le n, \tag{5.14}$$

where

p_i = strain measured after completing hole depth increment i,

P_j = uniform stress within hole depth increment j,

$\bar{a}_{ij}$ = strain response of a unit stress within increment j of a hole i increments deep, and

n = total number of hole depth increments.

In this discretization, the smoothly varying stress vs. depth profile represented by the dashed red line in Figure 5.6 is approximated by the stepped blue line shown, where the steps correspond to the hole depth increments used for the strain measurements. The stress within each hole depth increment is assumed to be uniform, thus producing the stepped pattern.

The relationship between the coefficients $\bar{a}_{ij}$ and the strain response function $\widehat{A}(H, h)$ is

$$\bar{a}_{ij} = \int_{H_{j-1}}^{H_j} \widehat{A}(H, h_i) \, dH. \tag{5.15}$$

The solution procedure for the stepped stresses P_j in Equation (5.14) is called the Integral Method. It proceeds in matrix format, where Equation (5.14) is expressed as

$$\bar{a} P = \frac{E}{1 + \nu} p. \tag{5.16}$$

In Equation (5.16), the order of the terms has been reversed to conform with conventional matrix notation. **Boldface** indicates matrix and vector quantities. For example, p is the vector of strains measured after each hole depth increment and P is the vector of stresses within each hole depth increment. Matrix $\bar{a}$ has a lower triangular structure, for example, for $n = 4$:

$$\bar{a} = \begin{bmatrix} \bar{a}_{11} & & & \\ \bar{a}_{21} & \bar{a}_{22} & & \\ \bar{a}_{31} & \bar{a}_{32} & \bar{a}_{33} & \\ \bar{a}_{41} & \bar{a}_{42} & \bar{a}_{43} & \bar{a}_{44} \end{bmatrix} \quad P = \begin{bmatrix} P_1 \\ P_2 \\ P_3 \\ P_4 \end{bmatrix} \quad p = \begin{bmatrix} p_1 \\ p_2 \\ p_3 \\ p_4 \end{bmatrix}. \tag{5.17}$$

The lower triangular matrix structure occurs because surface strains occur only in response to stresses within the hole depth. Deeper stresses have no influence. Thus, non-zero elements $\bar{a}_{ij}$ can appear in matrix $\bar{a}$ only when the stressed increment j is less or equal to the number of increments within the hole depth, i, i.e., when $1 \leq j \leq i \leq n$. Because of the triangular structure of $\bar{a}$, Equation (5.16) can be directly solved for the stress variation with depth P by forward substitution starting from the first row.

Figure 5.7 shows a physical interpretation of the elements $\bar{a}_{ij}$ of matrix $\bar{a}$. The rows of the matrix correspond to the strains caused by stresses within successive increments of a hole of given depth. The columns correspond to the strains caused by the stresses within a given depth increment, for holes of increasing depth. The summation of all the elements in each row corresponds to uniform stresses acting over the entire hole depth, as in Figure 5.2. Therefore, the row sums of matrix $\bar{a}$ equal the strain relaxation constant $\bar{a}$ in Equation (5.3) for uniform stresses in holes of the same total depth. The strain responses $\bar{a}_{ij}$ can be determined numerically using finite element calculations on the stress cases shown in Figure 5.7.

Similar remarks can be made about the required computations for the shear strains and stresses. Results analogous to Equations (5.13)–(5.17) can be written by substituting q and Q or t and T in place of p and P. In this case, all $\bar{a}$ quantities are replaced by analogous $\bar{b}$ quantities and the material-dependent term $E/(1 + \nu)$ is replaced by E. With these substitutions,

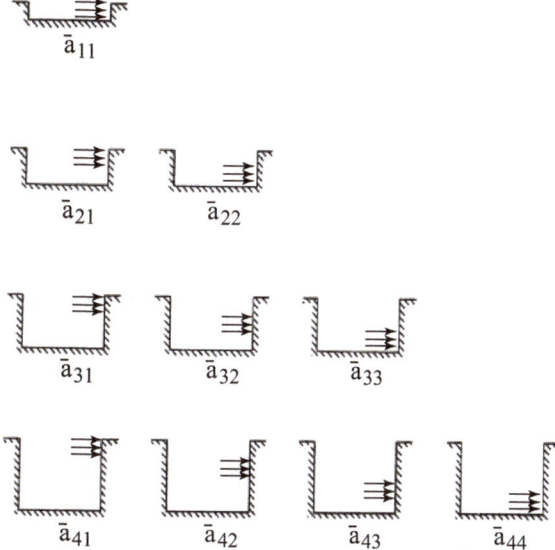

Figure 5.7: Schematic interpretation of the elements of matrix $\bar{a}$ (from Schajer (1988)).

Equation (5.16) generalizes to

$$\bar{a}P = \frac{E}{1+v}p \qquad \bar{b}Q = Eq \qquad \bar{b}T = Et.$$ (5.18)

In summary, the overall computational procedure is to start with the surface strains ε_1, ε_2 and ε_3 measured at a series of gradually increasing hole depths, convert them into vectors of the corresponding p, q and t strains using Equations (5.8), and then compute the corresponding P, Q and T stresses using Equations (5.18). These results can then be converted to axial or principal stresses using Equations (5.9)–(5.11). This procedure is directly analogous to the procedure described in Section 5.2 for calculating a single uniform stress. In this sense, the uniform residual stress case can be considered as a minimal example of an incremental hole-drilling measurement where the number of hole depth increments $n = 1$.

The matrix solution for stress in Equation (5.18) reveals a further significance of the use of the strain and stress quantities p, q, t and P, Q, T beyond a convenient grouping of terms in Equations (5.9)–(5.11). It can be seen that the three equations in (5.18) are entirely uncoupled and can be evaluated independently. The matrices $\bar{a}$ and $\bar{b}$ are of modest size, so the solution can proceed easily. In contrast, if the solution were instead done using axial strains and stresses, all three equations would need to be combined into a single, much larger equation with a matrix three times the size. The computation would require $3^3 = 27$ times more arithmetic operations, with correspondingly greater roundoff error effects.

5.6   PRACTICAL DETERMINATION OF $\bar{a}$ AND $\bar{b}$

The elements of matrices $\bar{a}$ and $\bar{b}$ can be determined using finite element computations that model the various loading cases shown in Figure 5.7. Since these computations are numerically intensive, it is convenient to do them just once for the case of standard equal hole depth increments and then to use the computed matrices as look-up tables for subsequent experimental work. ASTM Standard Test Method E837 specifies the use of 20 equally spaced hole depth increments and supplies the corresponding $\bar{a}$ and $\bar{b}$ numerical values for Types A, B and C strain gauge rosettes. The procedure described in E837 is recommended for standardized testing.

For specialized applications, it may be desirable to use different, possibly irregularly spaced hole depth increments. A potential approach to find the associated matrix values is to repeat the finite element calculations to fit the specific hole depth increments used. However, this can be a numerically intensive process. Instead, it is convenient to interpolate the linearly spaced results to fit the specific hole depth increments used. The separate use of matrices $\bar{a}$ and $\bar{b}$ in Equation (5.18) greatly assists in this process.

The arrangement in Figure 5.7 is unattractive for doing interpolations because it involves variations in three quantities: the hole depth and the depths at the upper and lower boundaries of the applied load. The number of variable quantities can be reduced to two by choosing to work instead with cumulative strain functions defined as

$$\bar{A}(H,h) = \int_0^H \widehat{A}(H,h)\, dH \qquad \bar{B}(H,h) = \int_0^H \widehat{B}(H,h)\, dH \tag{5.19}$$

in terms of which Equation (5.16) becomes

$$\bar{a}_{ij} = \bar{A}\left(H_j, h_i\right) - \bar{A}\left(H_{j-1}, h_i\right) \qquad \bar{b}_{ij} = \bar{B}\left(H_j, h_i\right) - \bar{B}\left(H_{j-1}, h_i\right). \tag{5.20}$$

Equation (5.20) shows that the use of the cumulative strain functions has replaced a three-variable interpolation with 2 two-variable interpolations and a subtraction. This process is straightforward to implement in practice because the same mathematical procedure can be used for both interpolations.

Figure 5.8 shows a schematic representation of the cumulative strain functions $\bar{A}(H,h)$ and $\bar{B}(H,h)$. Both have similar shapes, but with differences in local curvature and overall size. As with matrices $\bar{a}$ and $\bar{b}$, the cumulative strain functions are defined only in lower triangular regions. The inset diagram ② in Figure 5.8 gives an interpretation of the physical meaning of the cumulative strain functions. They correspond to the surface strains caused by a unit stress that extends to a depth H in a hole of depth h. These two depth quantities form the axes of the adjacent triangular diagram. Thus, a path in the 1–2 direction corresponds to the surface strain for increasing hole depth h, while keeping constant the depth limit H of the stress loading. Similarly, a path in the 2–3 direction corresponding to the surface strain for increasing depth limit H of the stress loading while keeping the hole depth h constant.

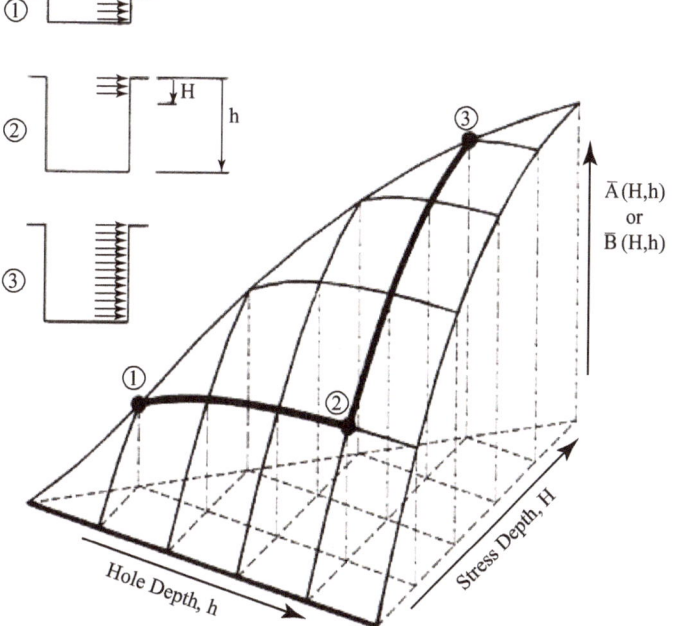

Figure 5.8: Schematic graphical representation of the cumulative strain functions $\bar{A}(H, h)$ and $\bar{B}(H, h)$ (from Schajer (1988)).

Table 5.2 illustrates an example tabulation of the cumulative strain functions for use with "Type A" strain gauge rosettes. To accommodate the three different available sizes, the hole depths quoted in the table can be scaled to fit each size, as described in the figure caption. Alternatively, in other published tables, the hole depth scaling may be done relative to the rosette mean radius or diameter.

Interpolations must be carried out with care because it is essential to reference only the numerical values within the lower triangle. Conventional rectangular interpolation schemes are unsatisfactory because they place zeroes in the upper triangle and then get false results by referencing those zeroes when doing interpolations near the diagonal.

Figure 5.9 shows a triangular interpolation scheme that constrains the interpolation within the lower triangle. This template should be placed within the tabulated numbers such that the target (H, h) coordinate, marked "X" in the diagram, lies within or as near as possible to the square area $BCDE$. Within the template, the target coordinate can be expressed in fractional format based on a local origin at C as (Y, y), where

$$H_X = H_C + Y \Delta H \qquad h_X = h_C + y \Delta h. \tag{5.21}$$

Note that the hole depth h increases downwards in Figure 5.9, so the ranges $-1 \leq Y \leq 0$ and $0 \leq y \leq 1$ exist within the area $BCDE$. Using a quadratic polynomial based on the six

Table 5.2: Cumulative strain function $\bar{A}(H, h)$ for a 1/16" Type A rosette with a 2 mm diameter hole. Depths h and H are in mm. Multiply depths by 0.5 for a 1/32" rosette or by 2 for a 1/8" rosette.

h	$\bar{A}$ (H,h)									
0.10	-0.0144									
0.20	-0.0189	-0.341								
0.30	-0.0224	-0.0416	-0.0563							
0.40	-0.0250	-0.0471	-0.0653	-0.0785						
0.50	-0.0269	-0.0510	-0.0716	-0.0880	-0.0993					
0.60	-0.0285	-0.0541	-0.0762	-0.0946	-0.1087	-0.1180				
0.70	-0.0296	-0.0563	-0.0797	-0.0993	-0.1151	-0.1267	-0.1341			
0.80	-0.0305	-0.0580	-0.0821	-0.1027	-0.1194	-0.1325	-0.1418	-0.1474		
0.90	-0.0310	-0.0592	-0.0839	-0.1050	-0.1225	-0.1363	-0.1468	-0.1541	-0.1582	
1.00	-0.0315	-0.0601	-0.0853	-0.1069	-0.1249	-0.1392	-0.1503	-0.1585	-0.1639	-0.1667
H→	0.10	0.20	0.30	0.40	0.50	0.60	0.70	0.80	0.90	1.00

tabulated points, the interpolated value at point X is

$$
\begin{aligned}
f_X = {} & y(y-1)/2 f_A + (1-y)(y-Y) f_B + (1-y)(1+Y) f_C \\
& + (y-Y)(y-Y-1)/2 f_D + (1+Y)(y-Y) f_E + Y(1+Y)/2 f_F,
\end{aligned} \qquad (5.22)
$$

where f_X, f_A, f_B, etc., are the $\bar{A}$ and $\bar{B}(H, h)$ values at points X, A, B, etc.

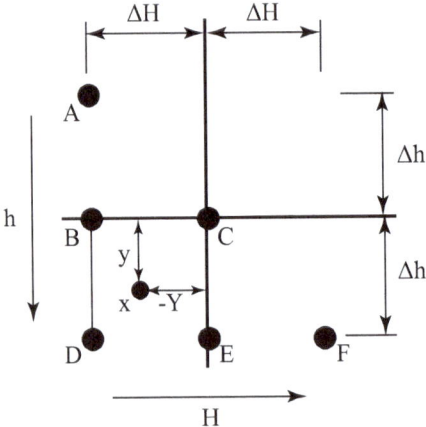

Figure 5.9: Template for interpolating the cumulative strain functions (from Schajer (1988)).

5.7 REGULARIZATION

After the numerical values of the $\bar{a}$ and $\bar{b}$ matrices have been determined either from tables or by interpolation, the residual stresses can be solved using Equations (5.18). Figure 5.10 shows the residual stresses computed from the strain measurements illustrated in Figure 5.5. It can be seen that the computed stresses are quite irregular and contain substantial noise, even though the measured strains on which they are based are fairly smooth with low noise. This type of noise amplification response is typical of the solution of an integral equation such as Equation (5.13). It is the mathematical manifestation of the Stress Effect described in Section 5.6, where the sensitivity of the measured surface strains to interior stresses is observed to diminish rapidly with distance from the surface.

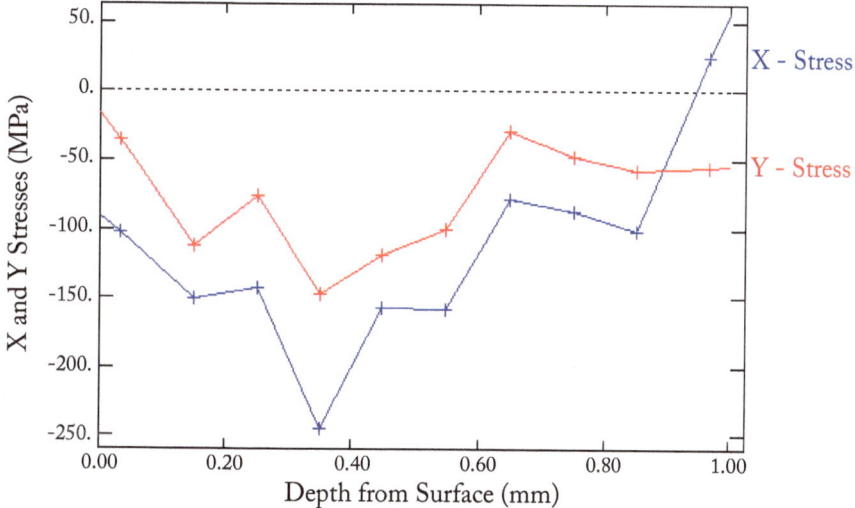

Figure 5.10: Unregularized stresses computed using Equations (5.18) from the strain data in Figure 5.5.

The diminishing sensitivity of the surface strains to the interior stresses can be seen numerically in matrices $\bar{a}$ and $\bar{b}$. Table 5.3 shows the matrix values that correspond to the hole depths and hole diameter used in the measurements shown in Figure 5.5. For simplicity here, they have been deliberately chosen also to correspond with the hole depths and diameter used in Table 5.2.

As can be seen from Table 5.3, the diminishing sensitivity to interior stresses causes the diagonal elements of matrix $\bar{a}$ rapidly to get smaller with increasing depth from the surface. The ratio of sizes of the diagonal element and leftmost elements diminishes row by row. By the last row, the ratio has dropped to 10%. If the table were continued to greater depths, this trend would accelerate, with values quickly approaching or even passing through zero. This behavior

Table 5.3: Matrix $\bar{a}$ for a 1/16" Type A rosette with a 2 mm diameter hole. Depths h and H are in mm. Multiply depths by 0.5 for a 1/32" rosette or by 2 for a 1/8" rosette.

h	[a]									
0.10	-0.0144									
0.20	-0.0189	-0.0152								
0.30	-0.0224	-0.0192	-0.0147							
0.40	-0.0250	-0.0222	-0.0182	-0.0132						
0.50	-0.0269	-0.0241	-0.0206	-0.0164	-0.0113					
0.60	-0.0285	-0.0256	-0.0221	-0.0184	-0.0141	-0.0093				
0.70	-0.0296	-0.0267	-0.0233	-0.0196	-0.0158	-0.0117	-0.0074			
0.80	-0.0305	-0.0276	-0.0241	-0.0205	-0.0168	-0.0131	-0.0093	-0.0056		
0.90	-0.0310	-0.0281	-0.0247	-0.0211	-0.0175	-0.0138	-0.0105	-0.0073	-0.0041	
1.00	-0.0315	-0.0286	-0.0252	-0.0216	-0.0179	-0.0144	-0.0111	-0.0082	-0.0054	-0.0028
H→	0.10	0.20	0.30	0.40	0.50	0.60	0.70	0.80	0.90	1.00

creates a serious numerical problem because the numerical conditioning of a matrix depends on the size of its determinant. For a triangular matrix, the determinant size is simply the product of all the diagonal elements. It is well known that if the determinant of a matrix is zero, no solution of an equation involving that matrix is possible. A near-zero determinant theoretically does allow a solution, but it is very ill-conditioned. Very small changes in any numerical value causes proportionally much larger changes in the computed solution. This is the effect seen in Figure 5.10 where small, barely perceptible noise in the measured strains causes substantial noise in the stress solution.

There are several possible strategies for reducing noise in the stress solution. The first, and most fundamental, is to use meticulous experimental technique so as to obtain the highest quality data containing absolute minimum of measurement noise. This is an essential requirement because no mathematical method can replace the information content missing from low-quality data. A second approach is to use only a small number of hole depth increments, preferably of increasing size. This strategy will leave fewer, larger numbers along the matrix diagonal, thereby increasing determinant size and numerical conditioning. A third possible approach is to smooth the measured strain data so as to reduce its noise content. This approach is effective, but care needs to be taken to avoid excessive smoothing or else the underlying data will be distorted. Smoothing also tends to reduce spatial resolution, so the ability to identify local details can easily be lost.

A further possible mathematical approach, now specified in ASTM Standard Test Method E837, is to use Tikhonov regularization to stabilize the residual stress solution. This computational technique is often used to improve the stability of inverse calculations. In concept, the approach seeks to separate the "true" part of the measured data, which it uses to determine the desired solution, from the "noise" content, which it seeks to reject. The mathematical procedure involves applying a penalty function on the stress solution that targets and diminishes the noise response while leaving the "true" response intact. This is done by observing that the desired stress solution is generally smooth, with few if any sharp features. In contrast, the noise contains many sharp features, indeed, that is its main characteristic. Thus, an effective noise rejection strategy is to seek to penalize and thereby diminish the presence of large curvatures within the computed stress solution. Mathematically, curvatures are indicated by the second derivative of the stress vs. depth profile, so the second derivative of the stress solution is the chosen penalty target.

The second derivative penalization is done by introducing a second term into the left sides of Equations (5.18) as follows:

$$\left(\bar{a}^T \bar{a} + \alpha_P c^T c\right) P = \frac{E}{1+\nu} \bar{a}^T p$$
$$\left(\bar{b}^T \bar{b} + \alpha_Q c^T c\right) Q = E \bar{b}^T q \qquad (5.23)$$
$$\left(\bar{b}^T \bar{b} + \alpha_T c^T c\right) T = E \bar{b}^T t.$$

Matrix c is an operator that acts on the stress solution to create a finite-difference evaluation of its local curvature. For equal hole depth increments, its structure is:

$$c = \begin{bmatrix} 0 & 0 & & & \\ -1 & 2 & -1 & & \\ & -1 & 2 & -1 & \\ & & -1 & 2 & -1 \\ & & & 0 & 0 \end{bmatrix}, \qquad (5.24)$$

where the number of rows equals the number of hole depth steps used. The first and last rows contain zeros, all other rows have second-derivative finite-difference operator $(-1\ 2\ -1)$ centered along the diagonal.

The factors α_P, α_Q and α_T control the amount of regularization that is applied. Zero values of the factors cause the regularized Equations (5.23) to reduce to the unregularized Equations (5.18). Positive values of the factors give regularization (smoothing) amounts that increase as larger factors are chosen. Excessively large regularization introduces so much smoothing that it distorts the underlying stress solution. Conversely, insufficient regularization allows excessive noise to remain in the calculated stress results. Optimal regularization balances these two tendencies, minimizing distortion of the stress solution while removing most noise.

Optimal regularization can be identified by considering the way in which Equations (5.23) affect strains. Solutions to the regularized Equations (5.23) give smooth stresses that differ from the theoretically "exact" but noisy stresses that would be calculated using the unregularized

Equations (5.18). Thus, if a circular calculation is done where the regularized stresses P, Q, T calculated from Equations (5.23) are substituted back into the unregularized Equations (5.18), the differences between the strain values corresponding to the smooth and noisy stresses correspond to the data distortion introduced by the regularization process. These differences are called "misfits."

$$p_{misfit} = p - ((1 + \nu)/E)\bar{a} P \quad q_{misfit} = q - (1/E)\bar{b} Q \quad t_{misfit} = t - (1/E)\bar{b} T. \qquad (5.25)$$

The Morozov criterion specifies how much misfit can be tolerated. The governing concept is that when the measured data are known only within a certain accuracy, there is no statistical justification for demanding a computation method to fit those data to a higher degree of accuracy. In this way, misfits up to the size of the experimental errors in the strain measurements can be tolerated because up to that level they are indistinguishable from noise. When using Equations (5.23) to compute the regularized stresses, it is therefore important to adjust the regularization factors α_P, α_Q and α_T such that the misfits reach but do not exceed the sizes of the respective measurement errors. These comparisons are done separately for each of the stresses P, Q and T.

The above process can proceed only if the sizes of the measurement errors are known. An order-of-magnitude estimate can be made quite quickly, but a reliable assessment requires a more detailed and time-consuming effort. It turns out that a surprisingly realistic estimate of measurement noise can be made by considering the deviations of the measured experimental data from a local best-fit curve. The standard deviation of the measurement noise can be estimated using

$$p^2{}_{std} = \sum_{j=1}^{n-3} \frac{(p_j - 3p_{j+1} + 3p_{j+2} - p_{j+3})^2}{20(n-3)} \qquad (5.26)$$

and similarly for the q and t strains.

When applied to the strain data in Figure 5.5, Equation (5.26) estimates each of the standard errors of the strain measurements to be approximately $0.9\ \mu\varepsilon$. The optimal regularization factors α_P, α_Q and α_T can then be determined iteratively so as to match the standard errors with the misfit values from Equations (5.25). Figure 5.11 shows the resulting regularized stresses that are computed using Equations (5.23). It can be see that the stress profiles are similar to the unregularized stresses in Figure 5.10, but with the noise removed. The corresponding strains back-calculated using Equations (5.18) are also indicated in Figure 5.5 as dashed lines. These are smoother than the solid lines for the measured strains, thereby showing that the regularization smooths both the computed stresses and the corresponding strains. The difference between the solid and dashed lines in Figure 5.5 is the "misfit." Its overall root-mean square average size is $0.9\ \mu\varepsilon$, corresponding to the target standard error for the regularization procedure.

The above stress computation procedure is adopted by the ASTM Standard Test Method E837. The Standard also provides tables of the required calibration matrices $\bar{a}$ and $\bar{b}$. Special-

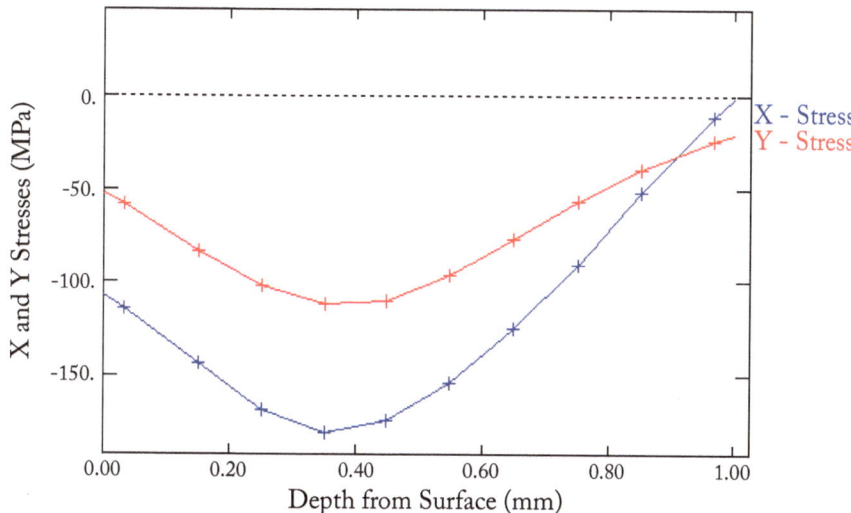

Figure 5.11: Regularized stresses computed using Equations (5.23) from the strain data in Figure 5.5. RMS misfit $= 0.9\ \mu\varepsilon$.

ized software for completing ASTM-style calculations is commercially available from various vendors.

5.8   OTHER CALCULATIONS

The technical literature on the Hole-Drilling Method spans a wide conceptual range. Many different methods have been proposed over the years to address various issues involved in computing residual stresses from hole-drilling data. The ASTM E837 procedure described above is the most widely accepted procedure and is recommended for general use. The following are some further procedures that may be useful in particular circumstances.

5.8.1   DIFFERENTIAL STRAIN AND AVERAGE STRESS METHODS

The Differential Strain and Average Stress methods were introduced during the early days of the development of the Hole-Drilling Method before finite element computations of calibration constants became generally available. At that time, calibration data were determined experimentally. However, it is not practically feasible to determine the elements of the calibration matrices $\bar{a}$ and $\bar{b}$ in this way. Both the Differential Strain and Average Stress methods were developed to compute the through-depth residual stress profile using experimental calibration data measured in hole-drilling tests on specimens containing known uniform residual stresses. In the Differential Strain Method, it is assumed that the surface strain change that occurs when the hole is drilled deeper by a small increment depends only on the stresses within that increment. Thus, the stresses within each increment can be determined by comparing the strain differences from one increment to the next with those from the calibration test. This approach ignores the geometric effect described in Section 5.6, and so produces only an approximate result. In terms of the Integral Method, the $\bar{a}$ and $\bar{b}$ matrices would have constant values down each column, arranged such that the sums of the rows are the same as those from the Integral Method. Because of its mathematical simplicity, the Differential Strain Method is still occasionally used for hole-drilling and ring-core calculations. However, because of the approximation involved, the results must be handled with caution.

In the Average Stress Method, it is assumed that the stresses at all depths contribute equally to the surface strain response. Thus, the stresses within a hole depth increment can be determined by finding the value that gives the same average as observed in the calibration test. This approach ignores the Stress Effect described in Section 5.6, so also produces only an approximate result. In terms of the Integral Method, the $\bar{a}$ and $\bar{b}$ matrices would be uniform within each row, arranged such that the sums of the rows are the same as those from the Integral Method. The degree of approximation of the Average Stress Method is greater than for the Differential Strain Method, and it is now rarely used.

5.8.2 POWER SERIES METHOD

The Power Series Method was introduced as an alternative to the Integral Method. Instead of representing the through-depth distribution of stresses in terms of the stress values within the various hole depth increments, it describes the stresses in terms of a power series based on depth from surface, h, i.e., $1, h, h^2 \ldots$, etc. While this approach is mathematically satisfactory, in practice, evaluations of the higher order terms tend to become unstable because of St. Venant effects. Consequently, only the lower terms can be determined reliably, typically just the first two terms: constant and linear. Because of this, the Power Series Method is suitable only for smoothly varying residual stress profiles. Going back one step further to the case of uniform stresses, the Power Series Method can be applied using just one term. In that case, it provides the Equivalent Uniform Stresses specified in Equations (5.12).

5.8.3 SPECIMEN THICKNESS

The ASTM Standard Test Method E837-13 recognizes two extreme measurement regimes: the "thick" case where the specimen thickness is greater than 1.0 times the strain gauge rosette mean diameter, and the "thin" case where the specimen thickness is less than 0.2 times the strain gauge rosette mean diameter. These numbers are for Type A and B rosettes; the corresponding factors for a Type C rosette are 1.2 and 0.24. The graphs of calibration constants in Figure 5.4 refer to the "thick" case. Most hole-drilling residual stress measurements are of the "thick" type.

The "thin" case occurs in measurements on sheet or plate type specimens, for which the hole goes all the way through the material. Historically, this configuration is important because it enabled the classical Kirsch analytical solution for the stress distribution around a hole in a stressed plate to be used as a basis for the computation of calibration constants. It is this solution that provides the factor $(1 + \nu)$ used to normalize calibration constant $\bar{a}$ in Equation (5.3). For very deep holes in "thick" specimens, the calibration constants tend toward the values from the Kirsch solution. If the curves in Figure 5.4 were extended further to the right, they would plateau at approximately the Kirsch values.

The Kirsch solution can be used directly for the plane stress case where the interior stresses are uniform through the material thickness. These are so-called "membrane" stresses. Out-of-plane constraints come into play if the plate thickness becomes appreciable compared with the hole diameter. Under these conditions, the plane stress assumption on which the Kirsch solution is based starts to break down, causing the solution to be less realistic. Table 3 in ASTM E837 includes calibration constants for use when working with "thin" plate specimens.

The intermediate case where the specimen thickness lies between the "thin" and "thick" specifications of E837-13 presents some challenges. The ASTM Standard Test Method suggests an intermediate behavior, but does not provide any explicit data. Studies of the "intermediate" case indicate a more complex response dominated by local bending around the drilled hole. Figure 5.12 illustrates the behavior. Such bending does not occur either in the "thin" case, where the through-hole creates a symmetrical geometry, or in the "thick" case, where the material is sufficiently massive to inhibit such deformation. However, in the "intermediate" case, the specimen geometry is neither symmetrical nor sufficiently massive. Since local bending occurs only in the intermediate case but not at either extreme, the intermediate thickness case cannot be handled through a simple interpolation between the "thin" and "thick" cases. Figure 5.13 illustrates the influence of material thickness on the calibration constants for four different hole depths. The similar values on the left of each graph correspond to the "thin" (through-hole) case. The values on the right of each graph correspond to the "thick" case. These latter values form an increasing sequence with hole depth, corresponding to the ascending curves in Figure 5.4. The sharp intermediate peaks in the $\bar{a}$ curves are caused by the bending effect illustrated in Figure 5.12. This bending is mostly axisymmetric, so it dominates calibration constant $\bar{a}$. Custom finite element calibrations are generally needed when working with intermediate thickness specimens.

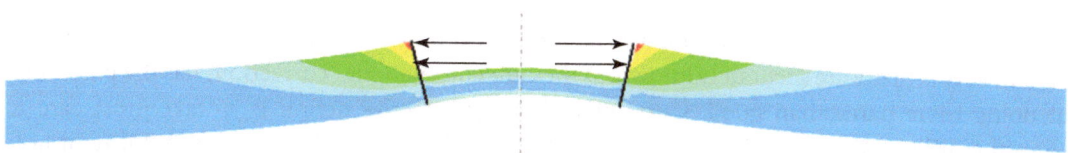

Figure 5.12: Localized bending caused by hole-drilling in an "intermediate" thickness specimen (from Schajer and Abraham (2014)).

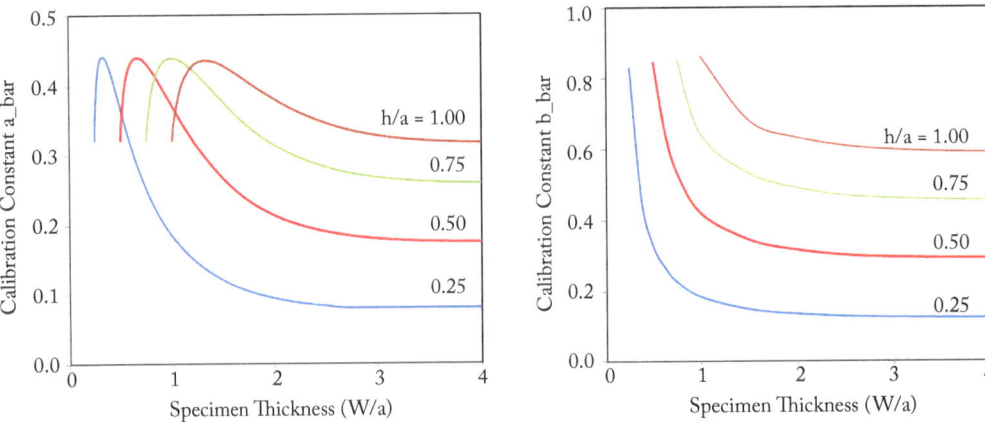

Figure 5.13: Variation with specimen thickness of the calibration constants $\bar{a}$ and $\bar{b}$ in an "intermediate" thickness specimen. h = hole depth, W = specimen thickness, a = hole radius (adapted from Schajer and Abraham (2014)).

5.8.4   HOLE ECCENTRICITY CORRECTION

The numerical values of the calibration matrices $\bar{a}$ and $\bar{b}$ are determined for the ideal case where the hole is exactly centered within the hole-drilling rosette used for the measurements, equally distant from each strain gauge. Commercial rosettes are made with prominent alignment marks and hole drilling devices include specially designed observation microscopes, both to assist with accurate hole placement. However, despite these precautions, it may sometimes happen that a hole is cut slightly off-center. In this case, the strain gauges are at unequal distances from the hole, with the nearer ones showing elevated readings and the more distant ones showing diminished readings. These strain variations distort the computed residual stresses.

The effect of hole eccentricity is purely geometrical, so if the size and position of the eccentricity are accurately known, the effect can be fully compensated mathematically. Practical methods have been proposed by Sandifer and Bowie (1978), Beghini (2010), and Ajovalasit (1979), among others. However, great care has to be taken when seeking to correct for hole eccentricity. The correction procedures can be effective only when the size and position of the eccentricity actually are accurately known. It they are not, then the "correction" has the possibility of doing more harm than good. Certainly, the most effective action is to take great care to align the drill accurately during equipment setup and thereby minimize the possibility of hole eccentricity errors.

5.8.5   PLASTICITY EFFECTS

When doing hole-drilling residual stress measurements, the drilling of the hole creates a stress concentration in the material around the hole. In a case where the original residual stresses are high, it may happen that the stress concentration effects may increase those stresses to reach the material yield stress, thereby causing localized plastic deformation around the hole. All stress evaluation equations quoted above apply specifically to the linear-elastic case and do not include any plasticity effects. When plasticity occurs, the additional plastic strains combine with the linear elastic strains to produce larger overall surface strains. This strain enlargement in turn produces an overestimation of the calculated residual stresses, potentially with values computed as being greater than the material yield stress.

The theoretical stress concentration factor produced around a through hole in a thin plate loaded by isotropic stresses corresponding to matrix $\bar{a}$ is 2, and for shear stresses corresponding to matrix $\bar{b}$ is 4. Fortunately, matrix $\bar{b}$ is not so poorly conditioned as matrix $\bar{a}$, so it does not respond so severely to its higher stress concentration factor. The material at the bottom of the blind hole used for hole-drilling measurements provides significant local reinforcement, significantly reducing the practical values of the stress concentration factors. In addition, loadings somewhat beyond those required to initiate plasticity at the hole edge are required to create significant plasticity in the strain gauge area. Taking all these factors together, the traditionally quoted size limit for the residual stresses that may be evaluated using the linear elastic formulas is 60% of the material yield stress. In more recent work, Beghini quotes 70% for a "full-depth" hole. Shallower holes have greater reinforcement from the material below the hole and so have lower stress concentration factors. With this is mind, the 2013 edition of ASTM E837 reduced the specified maximum hole depth and thereby enabled the allowance of a greater stress limit of 80% of the material yield stress.

Several authors, notably Beghini and coworkers sought to develop mathematical procedures to correct the computed stresses from hole-drilling measurements. These can enable realistic results to be achieved even when the residual stresses are very close to the material yield stress. Such corrections require accurate knowledge of both the material yield stress and of the stress/strain behavior immediately beyond the yield stress. These may not be easy to know reliably because the near-surface layer of many manufactured components is significantly harder than the bulk of the interior material.

Because of the nonlinearity caused by plasticity, more than three independent strain measurements are needed to evaluate the residual stresses in highly stressed materials. To meet this need, Beghini and coworkers proposed the use of the four-element strain gauge rosette shown in Figure 5.14, with the additional strain gauge 4 oriented perpendicular to the existing 45° gauge 2. Through the use of this rosette and the proposed mathematical correction procedures, hole-drilling measurements of residual stresses well beyond 90% of the material yield stress are reported to be possible.

5.8.6   ORTHOTROPIC MATERIALS

The computational framework described so far applies specifically to isotropic materials, i.e., materials whose physical properties are the same in all directions. This category covers the

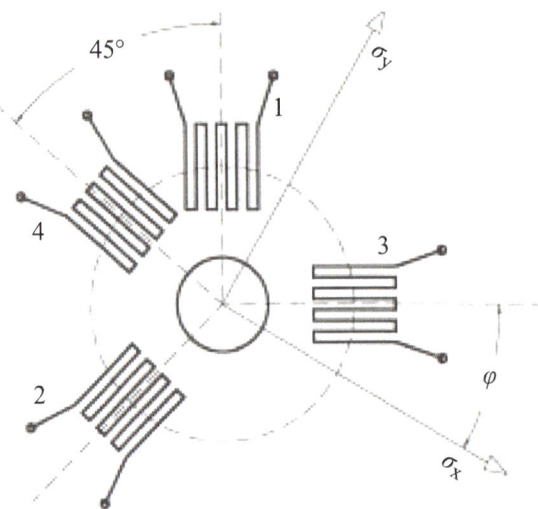

Figure 5.14: Proposed specialized strain gauge rosette design for residual stress measurements with plasticity correction (from Beghini (1998)).

large majority of engineering materials, but not all. An important minority of materials are orthotropic, i.e., they have different physical properties aligned with three mutually perpendicular axes within the material. Examples of such materials include fiber composites, wood, rolled metals, and many types of rock. Metals are generally elastically isotropic on the macro scale, but can be significantly orthotropic when considered on the scale of individual grains. This is particularly notable for hexagonal close packed metals, for example, in a single crystal of zinc the elastic constant in a basal direction is more than three times that in the prismatic direction.

The angular variation of strain in an orthotropic material is not trigonometric, so the foundational Equation (5.1) does not apply, neither most of the subsequent equations. In that case, a different approach needs to be taken. The strain response of an orthotropic material can be determined using finite element calculations using the approach described in Figure 5.2. Here, a full 3-D analysis is required because the previous use of two 2-D analyses based on isotropic and shear loading (involving $\bar{a}$ and $\bar{b}$) is no longer effective.

The in-plane residual stresses on an orthotropic material can be determined using a matrix-based formulation

$$
\frac{1}{\sqrt{E_x \, E_y}}
\begin{bmatrix}
c_{11} & c_{12} & c_{13} \\
c_{21} & c_{22} & c_{23} \\
c_{31} & c_{32} & c_{33}
\end{bmatrix}
\begin{bmatrix}
\sigma_x \\
\tau_{xy} \\
\sigma_y
\end{bmatrix}
=
\begin{bmatrix}
\varepsilon_1 \\
\varepsilon_2 \\
\varepsilon_3
\end{bmatrix}
\tag{5.27}
$$

where for the case where the ε_1 and ε_3 strain gauges in the standard hole-drilling rosette are aligned with the material elastic symmetry axes x and y, the coefficients c_{12} and c_{32} are zero.

There are then seven independent coefficients to be determined from the finite element analysis. This contrasts with the isotropic case, where there are only two independent coefficients. For the isotropic case, Equation (5.1) appears in matrix form as

$$
\begin{bmatrix} A+B & 0 & A-B \\ A & -2B & A \\ A-B & 0 & A+B \end{bmatrix} \begin{bmatrix} \sigma_x \\ \tau_{xy} \\ \sigma_y \end{bmatrix} = \begin{bmatrix} \varepsilon_1 \\ \varepsilon_2 \\ \varepsilon_3 \end{bmatrix}. \tag{5.28}
$$

Because of their greater complexity, residual stress measurements in orthotropic materials are generally limited to the basic case where the stress is uniform through the hole depth.

5.9   FURTHER READING

General References

- Mathar J (1934). Determination of Initial Stresses by Measuring the Deformation Around Drilled Holes. *Transactions ASME*, 56(4):249–254.

- ASTM (2013). Determining Residual Stresses by the Hole-drilling Strain-gage Method. *Standard Test Method E837-13*, American Society for Testing and Materials, West Conshohocken, PA.

- Grant PV, Lord JD, Whitehead PS (2002). The Measurement of Residual Stresses by the Incremental Hole Drilling Technique. *Measurement Good Practice Guide*, No. 53, National Physical Laboratory, Teddington, UK.

- Vishay Measurements Group, Inc. (1993). Measurement of Residual Stresses by the Hole-Drilling Strain-Gage Method. *Tech Note TN-503-6*, Vishay Measurements Group, Inc., Raleigh, NC. 16pp.

- Schajer GS and Whitehead P (2013). Hole-Drilling and Ring Core Methods. Chapter 2 in *Practical Residual Stress Measurement Methods*, G. Schajer, (Ed.), Wiley, Chichester, UK.

- Ajovalasit A, Scafidi M, Zuccarello B, Beghini M, Bertini L, Santus C, Valentini E, Benincasa A, Bertelli L (2010). The Hole-Drilling Strain Gauge Method for the Measurement of Uniform or Non-uniform Residual Stresses, *AIAS Residual Stress Working Group*, TR01:2010, Florence, Italy, 70pp.

- Oettel R (2000). The Determination of Uncertainties in Residual Stress Measurement (using the Hole Drilling Technique). *Code of Practice 15*, Issue 1, EU Project SMT4-CT97-2165.

- Rendler NJ, Vigness I (1966). Hole-Drilling Strain-gage Method of Measuring Residual Stresses. *Experimental Mechanics*, 6(12):577–586.

Calibration Method

- Beaney EM (1976). Accurate Measurement of Residual Stress on Any Steel Using the Center Hole Method. *Strain* 12(3):99–106.

- Schajer GS (1981). Application of Finite Element Calculations to Residual Stress Measurements. *Journal of Engineering Materials and Technology*, 103(2):157–163.

- Nau A, von Mirbach D, Scholtes B (2013). Improved Calibration Coefficients for the Hole-Drilling Method Considering the Influence of the Poisson Ratio. *Experimental Mechanics*, 53(8):1371–1381.

- Boiten RG, ten Cate W (1952). A Routine Method for the Measurement of Residual Stresses in Plates. *Applied Scientific Research* A3(5):317–343.

- Bueckner H (1958). The Propagation of Cracks and the Energy of Elastic Deformation. *Transactions ASME*, 80:1225–1230.

- Baldi A (2017). Far-Field Boundary Conditions for Calculation of Hole-Drilling Residual Stress Calibration Coefficients. *Experimental Mechanics*, 57(4):659–664.

- Blödorn R, Bonomo LA, Viotti MR, Schroeter RB, Albertazzi A (2017). Calibration Coefficients Determination Through FEM Simulations for the Hole-Drilling Method Considering the Real Hole Geometry. *Experimental Techniques*, 41(1):37–44.

Strain Gauge Modeling

- Schajer GS (1993). Use of Displacement Data to Calculate Strain Gauge Response in Non-uniform Strain Fields, *Strain*, 29(1):9–13.

Incremental Stress Computation

- Bijak-Zochowski M (1978). A Semidestructive Method of Measuring Residual Stresses. *VDI-Berichte*, 313:469–476.

- Niku-Lari A, Lu J, Flavenot JF (1985). Measurement of Residual-stress Distribution by the Incremental Hole-Drilling Method, *Experimental Mechanics*, 25(2):175–185.

- Schajer GS (1988). Measurement of Non-uniform Residual Stresses Using the Hole-Drilling Method. *Journal of Engineering Materials and Technology*, 110(4) Part I:338–343, Part II:344–349.

- Zuccarello B (1999). Optimal Calculation Steps for the Evaluation of Residual Stress by the Incremental Hole Drilling Method. *Experimental Mechanics*, 39(2):117–124.

- Schajer GS (2007). Hole-Drilling Residual Stress Profiling with Automated Smoothing. *Journal of Engineering Materials and Technology*, 129(3):440–445.

- Schajer GS, Rickert TJ (2011). Incremental Computation Technique for Residual Stress Calculations Using the Integral Method. *Experimental Mechanics*, 51(7):1217–1222.

- Schajer GS, Winiarski B, Withers PJ (2013). Hole-Drilling Residual Stress Measurement with Artifact Correction Using Full-field DIC. *Experimental Mechanics*, 53(2):255–265.

Alternative Stress Calculation Methods

- Kelsey RA (1956). Measuring Non-uniform Residual Stresses by the Hole Drilling Method. *Proc. SESA*, 14(1):181–194.

- Nickola WE (1986). Practical Subsurface Residual Stress Evaluation by the Hole Drilling Method, *Proc. SEM Spring Conference on Experimental Mechanics*, New Orleans, pp. 47–58.

- Schajer GS (1991). Strain Data Averaging for the Hole-Drilling Method. *Experimental Techniques*, 15(2):25–28.

Error Analysis

- Vangi D (1994). Data Management for the Evaluation of Residual Stresses by the Incremental Hole-Drilling Method. *Journal of Engineering Materials and Technology*, 116(4):561–566.

- Schajer GS, Altus E (1996). Stress Calculation Error Analysis for Incremental Hole Drilling Residual Stress Measurements. *Journal of Engineering Materials and Technology*, 118(1):120–126.

- Scafidi M, Valentini E, Zuccarello B (2011). Error and Uncertainty Analysis of the Residual Stresses Computed by Using the Hole Drilling Method. *Strain*, 47(4):301–312.

- Casavola C, Pappalettera G, Pappalettere C, Tursi F (2013). Analysis of the Effects of Strain Measurement Errors on Residual Stresses Measured by Incremental Hole-Drilling Method. *The Journal of Strain Analysis for Engineering Design*, 48(5):313–320.

Thin Plate Specimen

- Schajer GS, Abraham C (2014). Residual Stress Measurements in Finite-thickness Materials by Hole-drilling. *Experimental Mechanics*, 54(9):1515–1522.

- Held E, Schuster S, Gibmeier J (2014). Incremental Hole-Drilling Method vs. Thin Components: A Simple Correction Approach. *Advanced Materials Research*, 996:283–288.

- Schuster S, Steinzig M, Gibmeier J (2017). Incremental Hole Drilling for Residual Stress Analysis of Thin Walled Components with Regard to Plasticity Effects. *Experimental Mechanics*, 11pp. https://doi.org/10.1007/s11340-017-0318-7

Hole Eccentricity Compensation

- Sandifer JP, Bowie GE (1978). Residual Stress by Blind-hole Method with Off-center Hole. *Experimental Mechanics*, 18(5):173–179.

- Ajovalasit A (1979). Measurement of Residual Stresses by the Hole-Drilling Method: Influence of Hole Eccentricity. *The Journal of Strain Analysis for Engineering Design*, 14(4):171–178.

- Beghini M, Bertini L, Mori LF (2010). Evaluating Non-uniform Residual Stress by the Hole Drilling Method with Concentric and Eccentric Holes. Part I and Part II, *Strain*, 46(4):324–336 and 46(4):337–346.

- Nau A, Scholtes B (2012). Experimental and Numerical Strategies to Consider Hole Eccentricity for Residual Stress Measurement with the Hole Drilling Method. *Materials Testing*, 54(5):296–303.

Material Plasticity Compensation

- Beghini M, Bertini L, Raffaelli P (1995). An account of Plasticity in the Hole-Drilling Method of Residual Stress Measurement. *The Journal of Strain Analysis for Engineering Design*, 30(3):227–233.

- Beghini M, Bertini L (1998). Recent Advances in the Hole Drilling Method for Residual Stress Measurement. *Journal of Materials Engineering and Performance*, 7(2):163–172.

- Beghini M, Bertini L, Santus C (2010). A Procedure for Evaluating High Residual Stresses by the Blind Hole Drilling Method Including the Effect of Plasticity. *Journal of Strain Analysis for Engineering Design*, 45(4):301–318.

Orthotropic Materials

- Schajer GS, Yang L (1994). Residual-stress Measurement in Orthotropic Materials using the Hole-Drilling Method. *Experimental Mechanics*, 34(4):324–333.

- Pagliaro P, Zuccarello B (2007). Residual Stress Analysis of Orthotropic Materials by the Through-hole Drilling Method. *Experimental Mechanics*, 47(2):217–236.

- Prasad CB, Prabhakaran R, Thompkins S (1987). Determination of Calibration Constants for the Hole-Drilling Residual Stress Measurement Technique Applied to Orthotropic Composites. *Composite Structures, Part I: Theoretical Considerations*, 8(2):105–118; Part II: Experimental Evaluations 8(3):165–172.

Residual Stress Computation Software

- Schajer GS. H-drill—Hole-Drilling Residual Stress Calculation Program. `http://www.schajer.org/index.htm`

- SINT Technology. Restan MTS3000 Measurement System. `http://www.residualstressmeasurement.com/the_hole-drilling_method.html`

- Nobre JP, Dias AM, Domingos AJ, Morais R, Reis MJCS (2009). A Windows-based Software Package to Evaluate Residual Stresses by the Incremental Hole-Drilling Technique. *Computer Applications in Engineering Education*, 17(3):351–362. `https://www.researchgate.net/journal/1061-3773_Computer_Applications_in_Engineering_Education`

CHAPTER 6

Example Practical Procedures and Results

"Operator skill has been identified as probably the most important factor in achieving a reliable and quality measurement."

National Physical Laboratory Measurement Good Practice Guide No. 53 (2006).

This chapter describes practical examples of strain gauge installations to illustrate further aspects of the Hole-Drilling Method. Sections 6.1–6.5 describe examples of rosette installations on specimens in practical engineering situations. Sections 6.6–6.10 go on to present and discuss further examples of residual stress measurements on a range of different types of specimens.

6.1 SPECIMEN GEOMETRY AND STRAIN GAUGE SELECTION DETAILS

ASTM Standard Test Method E837 for hole-drilling describes the required characteristics of an ideal test specimen. The measurement location on such a specimen has a plane smooth surface that is remote from any obstacles, edges, or discontinuities such as holes or steps. In addition, the specimen material is linear-elastic, isotropic and homogeneous.

It is convenient to express the dimensions of specimens as multiples of the mean gauge diameter D. This is the pitch diameter of the gauge elements set around the target center. The ideal specimen surface is flat, but curvatures with radii down to ~2.4D can be accommodated with little impact on the results. Another important property of the specimen is the material thickness in the area of measurement. For specimens to be subjected to through-hole drilling measurements, ASTM E837 lists the maximum thickness as 0.2D for Type A and Type B gauges and 0.24D for the Type C gauge. For blind hole (incremental) drilling, the minimum thickness is 1.0D for Type A and Type B gauges and 1.2D for the Type C gauge.

Figure 6.1 shows the layouts and key dimensions of the widely used Type A and Type B strain gauge rosettes. Table 6.1 summarizes the geometrical requirements for specimen thickness, distance from adjacent features and surface shape. The main specifications in the table derive from the ASTM E837, with some additional suggestions based on the practical experience of the authors.

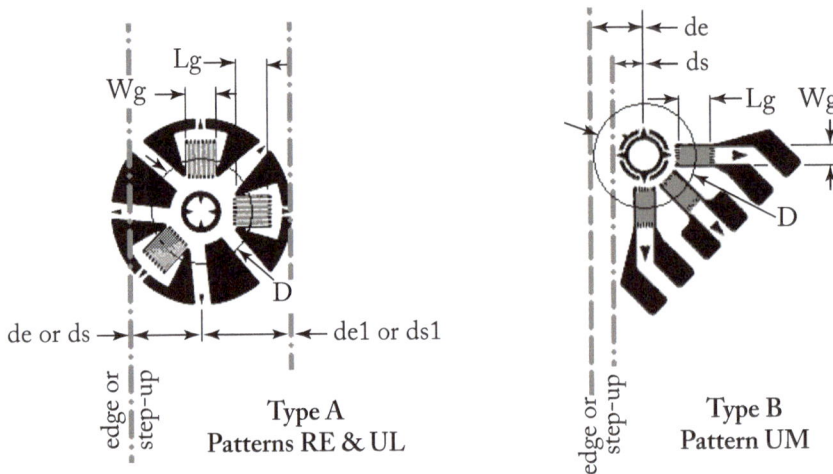

Figure 6.1: Target strain gauge rosette layouts and key dimensions.

Key dimensions are:

D = mean gauge diameter

D_o = nominal hole diameter

Lg = length of gauge element

Wg = width of gauge element

de or $de1$ = distance to edge feature

ds or $ds1$ = distance to step-up feature

Dh = drilled hole diameter

$zs_{\max}$ = maximum stress depth

The gauge size is the principal factor to be considered for measurement planning because it determines the maximum depth to which residual stresses can be detected. However, the maximum gauge size may be limited by the specimen thickness and the proximity of the proposed gauge location to any nearby specimen features.

The three types of ASTM strain gauge rosettes identified in Figure 4.1 (Chapter 4) are manufactured in several sizes and configurations. Type A rosettes are manufactured in 1/32", 1/16", and 1/8" nominal sizes as "open" gauges (with an exposed metal surface) and in the 1/16" size as an encapsulated gauge (with the metal surface covered by a protective film) with large copper solder pads. Gauge leads can be soldered directly to the solder pads of encapsulated

Table 6.1: Geometrical specifications for hole-drilling rosette use, adapted and expanded from ASTM E837. Unless otherwise stated, dimensions are mm. (* indicates authors' suggestions.)

	Symbol	ASTM E837 Rosette Type					
		A				B	C
Vishay Pattern	-	031RE	062RE	062UL	125RE	062UM	030RR
Nominal element size	Inches	1/32	1/16	1/16	1/8	1/16	-
Gauge mean diameter	D	2.57	5.13	5.13	10.26	5.13	4.32
Gauge length	L_g	0.8	1.6	1.6	3.2	1.6	~0.8
Gauge width	W_g	0.8	1.6	1.6	3.2	1.2	~0.8
Nominal hole diameter	D_o	1.0	2.0	2.0	4.0	2.0	2.0
Maximum drilled hole depth	zh_{max}	0.5	1.0	1.0	2.0	1.0	1.25
Minimum specimen thickness	ts_{min}	3.0	6.0	6.0	12.0	6.0	8.0
Min. distance to edge feature	de_{min}	2.5	5.0	5.0	10.0	4.0	5.0
Min. distance to edge feature	$de1_{min}$	2.8	5.7	5.7	11.4	-	5.0
Min. distance to step feature*	ds_{min}	2.0	4.0	4.0	8.0	2.0	5.0
Min. distance to step feature*	$ds1_{min}$	2.8	5.7	5.7	11.4	-	5.0
Min. radius of curvature*	rc_{min}	6.0	12.0	12.0	24.0	12.0	12.0
Min. gauge-to-gauge distance*	dg_{min}	6.0	12.0	12.0	24.0	12.0	12.0

gauges, while open gauges require the use of intermediate conductors soldered between the rosette and remote terminal pads to which the gauge leads are soldered. The Type B rosette is produced as an encapsulated 1/16" gauge while the Type C rosette is produced as an open 030-size gauge and is identified as pattern RR. Similar rosette patterns are produced by various manufacturers and are all suitable for use providing their specific calibration data are used for the residual stress calculation step.

The encapsulated type of strain gauge rosette is more robust and less prone to damage during installation and soldering. Thus, it is advantageous for general-purpose use. However, the increase in thickness can result in excessive bond thickness when bonding to an irregular or significantly curved specimen surface. In that case, the flexibility of an "open" rosette makes it a preferable choice.

6.2   PRACTICAL STRAIN GAUGE ROSETTE INSTALLATIONS

Practical specimens are often less than fully ideal because of their complex geometry and the presence of local geometric features. The photographs in Figures 6.2–6.6 illustrate some of the dimensions defined in Table 6.1 and show ways in which the standard rosette patterns can be used and possibly adapted to accommodate various practical circumstances. The extreme gauge position dimensions (minimum or maximum) are listed in Table 6.1 for which the published Integral Method coefficient values remain valid within a range of ca. ±4%. These dimensions relate to the:

- specimen thickness (ts),

- distance to a free edge (de and de1),

- distance to step features (ds and ds1),

- radius of curvature of the surface (rc).

In non-standard examples of gauge installations, coefficients for combinations of extreme installation dimensions (for example near-edge and small radius of curvature) may lie further from the ASTM E837 values.

Dimensions relating to the drilled hole, specimen, and gauge position are established as follows.

- *Nominal hole diameter*: Typically, $D_o = 0.4D$ with a recommended range of diameters of ±0.04D. The use of a smaller hole diameter is a concern because it leads to lower strain outputs, in particular at depths close to the surface. This reduced strain response increases uncertainties in the calculated residual stress values. Conversely, the use of a larger hole diameter is a concern because it can lead to gauge and bond damage close to the innermost parts of the gauge elements. Accordingly, for 1/16"-size gauges ($D = 5.13$ mm), a good choice for the nominal hole diameter is around 2.0–2.2 mm.

- *Drilled hole depth*: The maximum depth zh_{max} to which holes are drilled is defined by the size and geometry of the gauge pattern. For rosette Types A and B, ASTM E837 specifies $20 \times 0.02D$ drilling steps for a "uniform" stress measurement (final depth of 0.4D) and $20 \times 0.01D$ for an incremental measurement (final depth 0.2D). The smaller depth allows for measurements closer to the material yield stress. The specified drilling steps for the Type C rosette for an incremental measurement is $25 \times 0.01D$ (final depth 0.25D). Drilling

 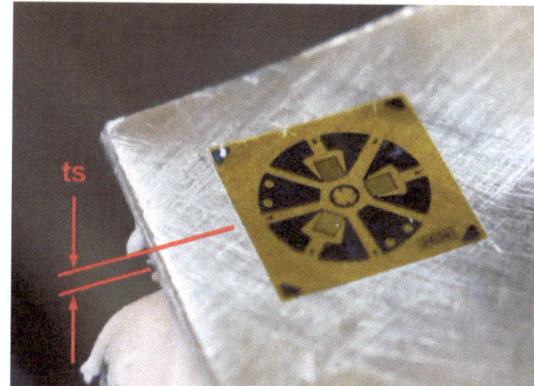

(a) Specimen thickness (moderate)   (b) Specimen thickness (thin)

Figure 6.2: Gauge rosette installations on thin specimens (photos courtesy of Stresscraft Ltd.).

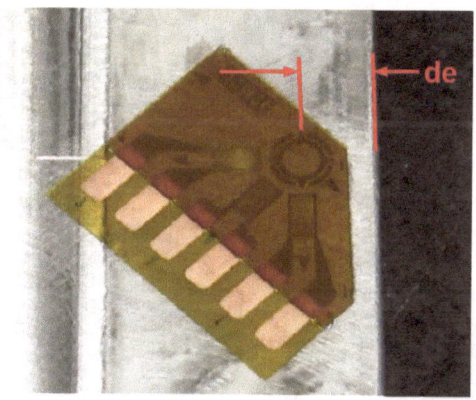 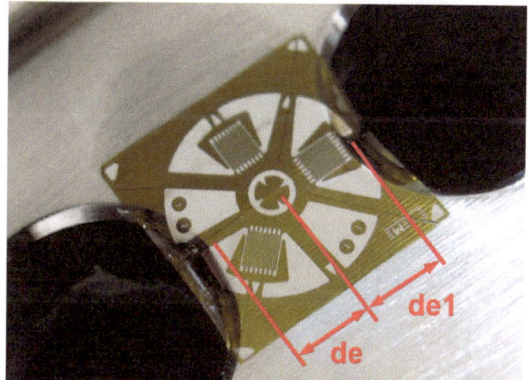

(a) Distance to a single free edge   (b) Distances to two free edges

Figure 6.3: Gauge rosette installations adjacent to edge features (photos courtesy of Stresscraft Ltd.).

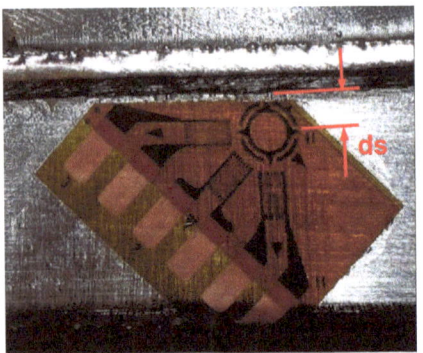

(a) Distance to a single step feature

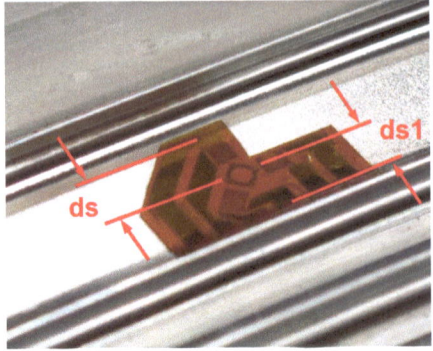

(b) Distances to two step features

Figure 6.4: Gauge rosette installations adjacent to step features (photos courtesy of Stresscraft Ltd.).

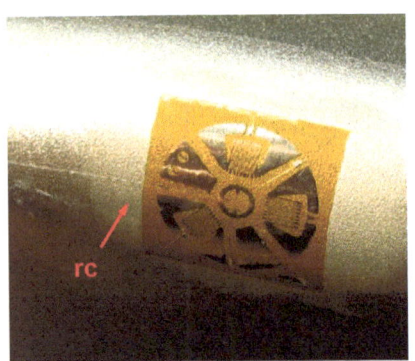

(a) Surface curvature (external, single)

(b) Surface curvature (external, double)

(c) Surface curvature (internal, single)

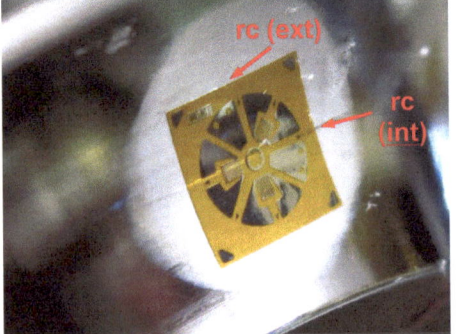

(d) Surface curvature (external and external)

Figure 6.5: Gauge rosette installations on curved surfaces (photos courtesy of Stresscraft Ltd.).

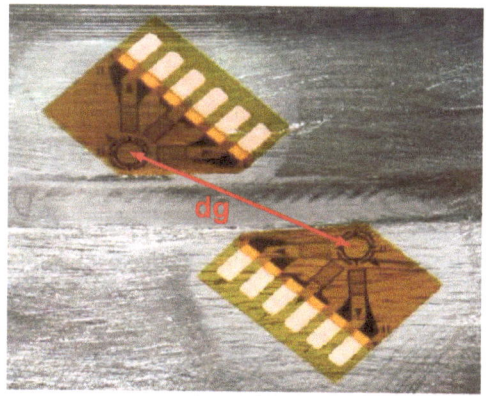
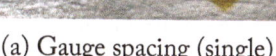

(a) Gauge spacing (single)

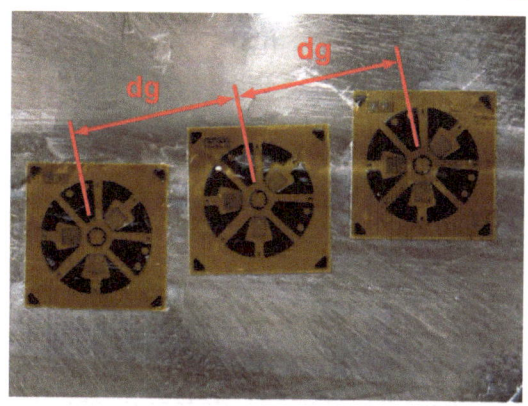

(b) Gauge spacing (multiple)

Figure 6.6: Multiple strain gauge rosette installations (photos courtesy of Stresscraft Ltd.).

beyond these depths produces no useful further information about interior stresses. The maximum depth to which residual stresses can be evaluated is defined by the size and geometry of the gauge pattern. This depth is slightly smaller than $zh_{\max}$; because the computed stress within each hole depth increment is associated with the center of the increment, the effective maximum stress data depth is reduced below the hole depth by half the thickness of the final calculation increment.

- *Specimen thickness*: For measurements on a thin specimen, the drilling process causes significant localized bending of the specimen. For rosette Types A and B, ASTM E837 proposes a minimum specimen thickness $ts_{\min} = 1.0D$. Rosette Type C includes circumferential elements for which deeper hole depths cause significantly greater changes in stiffness. Accordingly, the minimum specimen thickness for Type C rosettes is somewhat greater (around 1.2D). Figures 6.2a and 6.2b show 1/32" rosettes installed on specimens where the material thickness is to be considered; the thinner specimen is shown in Figure 6.1b.

- *Distance to an edge feature*: The proximity of a free edge to the drilled hole causes changes in stiffness and departures from the conditions used to calculate the published Integral Method coefficients. This is a particular concern when using rosette Type B (Figure 6.3a), where all the gauge elements are contained within a single quadrant. In extreme cases, the drilled hole could be positioned very close to an edge. Where the distance between the gauge center and specimen edge $de_{\min}$ is less than 0.8D, the validity of published Integral Method coefficients may lie outside the range $\pm4\%$. The presence of chamfers at the edge may further influence the validity of the coefficients. Figure 6.3b shows a Type A rosette installed between two chamfered holes; in this case the close proximity of the gauge elements to the free edges may result in departure from the conditions required for the published coefficients.

- *Distance to a step feature*: The proximity of a step feature in the region of the gauge results in changes to the stiffness of material around the drilled hole that are usually less severe than changes caused by an edge feature at the same distance. It is unlikely that the presence of small steps, such as weld beads (Figure 6.4a), will affect the validity of published Integral Method coefficients for practical ranges of dimension "ds" that can be achieved using available rosettes. The two adjacent steps shown in Figure 6.4b may also have an effect on the gauge outputs, but the presence of blend fillets will reduce the severity of this. For both of these applications, parts of the rosette backing material can be removed to increase the flexibility of the rosette for ease of installation. Distances from drilled holes to larger steps should be greater than the tabulated values to avoid excessive uncertainties.

- *Radius of curvature*: Figures 6.5a–d show the installation of gauges on curved surfaces. In Figure 6.5a, an open 1/16" pattern gauge rosette is installed on an external cylindrical surface; Figure 6.5b shows a 1/8" rosette installed on a spherical surface. The rosette in Figure 6.5c is installed on a fillet surface while the gauge in Figure 6.5d is installed on a "saddle" surface with both internal and external curvatures.

  Curvature of the gauge installation surface causes two significant concerns:

  – the required Integral Method coefficients for a curved surface (cylindrical or spherical) will progressively deviate from the published values for a flat surface as curvature increases; and

  – drilling a flat-bottomed hole into a curved surface results in ambiguity concerning the selection of the hole datum depth (from which all subsequent depth increments are measured). This uncertainty is most significant at shallow drilling increments at surfaces with a small radius of curvature.

In practice, where the radius of curvature falls below the quoted rc_{min} value, the effect on coefficients and shallow increment strains can be investigated using finite element models and specimens with "known" stresses from applied loads. Figure 6.5b also shows how slitting of the rosette backing material between the adjacent gauge elements can be used to provide the required flexibility of the rosette to wrap the elements around the curved specimen surface. A less severe gauge installation is shown in Figure 6.5c where much of the backing material of the encapsulated 1/16"-size rosette has been removed to increase the flexibility of the gauge to facilitate wrapping around the fillet.

- *Gauge spacing:* The relaxation effects of hole drilling extend beyond the boundaries of the rosette. Where it is required to install a number of rosettes on a specimen surface (Figures 6.6a and 6.6b) potential interference between adjacent holes must be considered. For a hole of diameter 0.4D drilled to a depth of 0.2D in an equi-biaxial stress field, the stress relaxation at a distance of 2.4D is less than 1% of the stress originally at the hole. Accordingly, it is recommended that the minimum distance between adjacent holes dg_{min} should be at least 2.4 times the gauge diameter D.

Some further factors that need to be considered when planning measurements include the following.

- **Drilling machine accuracy.** The Hole-Drilling Method requires making strain measurements after each of a sequence of hole depth increments. The accuracy of the subsequent residual stress evaluation directly depends on the dimensional accuracy of the hole depth increments, which in turn depends on the quality of the cutter position control during the drilling process. This issue becomes more of a concern when using many small hole depth increments (10–25 are typical) and when using very small rosettes. In the latter case the required hole diameter and depth increments vary in proportion to the rosette size, and so a greater absolute accuracy is required to maintain a reasonable level of relative accuracy of the drill depth increments and the concentricity of the drill within the rosette. Thus, while the smallest 1/32" rosettes may appear attractive to fit in sites with restricted area or on thin, curved specimens, they can be less practical because of their more demanding drilling accuracy, handling and soldering requirements.

- **Number of gauges to be installed on the specimen.** As noted above, the stress relaxation effects of hole-drilling extend over areas of the specimen beyond the immediate area covered by the target gauge. Accordingly, where it is required to install and drill a large number of gauges within a restricted surface area, the limit on between-hole distance dg_{min} specified in Table 6.1 can be met by using smaller rosettes, but with the caution noted above. In addition, where a mix of large and small gauge sizes are to be applied to a specimen in close proximity to each other, planning for the sequence of measurements should consider the installation and drilling of the smaller gauges (with smaller diameter holes) prior to the larger gauges.

Where the installation of the gauge on the specimen surface differs significantly from the above specifications, detailed finite element models of the specimen (incorporating the drilled hole) can be used to quantify the strain response of the residual stresses. However, useful comparative information can often still be gained from processing non-compliant strain data using the standard Integral Method coefficients. For example, comparison of results from two or more differently processed specimens can indicate large differences in residual stresses even though the results are not strictly quantitative.

6.3 ORIENTATION OF TYPE B STRAIN GAUGE ROSETTES

The one-sided arrangement of Type B rosettes enables hole-drilling measurements to be made on specimens at locations close to edge and step features. However, the use of such gauges requires careful consideration prior to installation to achieve the best possible results. The layout of gauge elements and row of solder tabs appears to invite the user to install such a gauge with the "long" side of the rosette parallel to the edge or step feature. This direction commonly corresponds to a principal strain direction. Figure 6.7a shows such a direct installation. In this case, the minimum principal strain is perpendicular to the edge and the maximum principal strain is parallel to the edge. This arrangement is non-ideal because only one strain gauge (number 2) is aligned in a principal direction. The strain required for the computation of the stress σ_{max} in Figure 6.7a can be calculated from combining the strains from gauge elements 1, 2, and 3. However, consideration of the Mohr's circle for this "worst" alignment case demonstrates that the resulting uncertainty in σ_{max} can be up to 3 times that found when the principal strains are aligned with gauges 1 and 3.

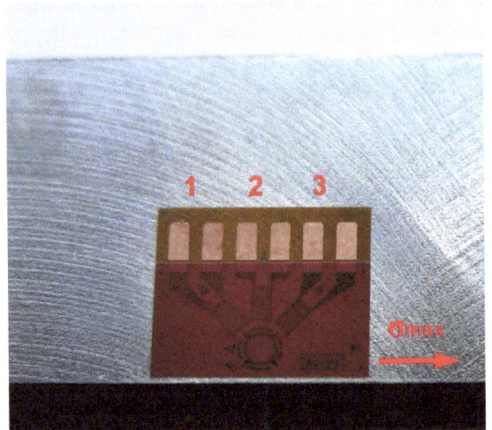

(a) Type B gauge aligned with edge   (b) Type B gauge; inclined installation

Figure 6.7: Type B gauge at an edge feature; direct installation (images courtesy of Stresscraft Ltd.).

In practical terms, such an amplification of uncertainty can add an unacceptable level of noise to the subsequent stress computation. This problem can be avoided by trimming some of the gauge backing material and re-orientating the rosette so that one of the gauge elements is aligned with the edge feature. Figure 6.7b shows such an installation; for this revised installation, uncertainties in both the principal strain directions are minimized.

This type of orientation has been used for all Type B rosette installations shown in the previous sections in this chapter. It may be the case that, at some features, the directions of principal strains and stresses are not known prior to the test (for example, from additional shear loading along a weld line). In these cases, a preliminary gauge can be installed and drilled to establish the directions of principal stresses prior to installing rosette(s) at some distance from the preliminary gauge.

6.4   INSTALLATION ON IRREGULAR SURFACES: BOND THICKNESS

Chapter 4 discusses specimen surface preparation for gauge installation. For measurements on irregular surfaces where it is deemed unacceptable to apply any smoothing, additional measures must be taken to ensure that the bond between the rosette and specimen is as thin as possible. To achieve such a bond, the most flexible gauge construction must be selected, for example an open 1/32" or 1/16"-size rosette. Figure 6.8a shows such a rosette installed on an unsmoothed cast surface, illuminated by a uniform, low-level light source. It may not be possible to determine whether a suitable bond has been achieved by visual inspection under these conditions. Figure 6.8b shows a rosette installed on a shot-peened surface, illuminated by a bright spotlight source to create clear reflections of the light on the rosette surface. By inspection of the surface in this way, it is possible to form a judgement as to whether the gauge backing material follows the surface of the specimen. Viewing under the conditions in Figure 6.8b is also useful for determining whether any areas of the adhesive have cured prematurely, prior to application of the bond pressure.

Figure 6.8c provides another example of viewing the gauge with spot-light illumination. In this case, the specimen surface is a machined (turned) surface with very much smaller irregularities than the shot-peened surface in the previous example. The "witness" of the machining marks can be clearly seen in the reflected light. By applying high pressure during bonding, it is possible to achieve very thin bond lines such that quite fine surface markings become visible at the open gauge rosette surface.

A more quantitative approach to bond thickness assessment can be achieved using the arrangement shown in Figure 6.8d. In this example, a section through an open gauge, the rosette surface is illuminated by a point light source at an angle of ~45°. The user then views the gauge in a direction perpendicular to the target surface using the drilling machine optical head. Any offset in the plane of the surface between the outline of the metallic gauge film and its shadow when viewed on the target surface through the transparent backing material and adhesive equals the height of the gauge film above the surface. The offset can be measured using the optical head graticule scale. Subtraction of the backing material thickness from the shadow offset provides an indication of the bond thickness. This thickness is typically around 20–30 μm. The same method can be used to determine any variability in bond thickness under encapsulated gauges, where inspection of the encapsulation surface is less revealing than for open gauges.

(a) Open rosette on a cast surface

(b) Open rosette on a shot-peened surface

(c) Open rosette on a machined surface

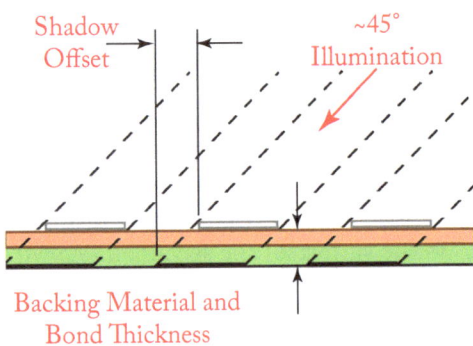

(d) Bond thickness measurement

Figure 6.8: Strain gauge rosette installations on irregular surfaces (images courtesy of Stresscraft Ltd.).

6.5 NON-STANDARD GAUGES

In some instances, specimen conditions or strain gauge requirements may arise for which no standard strain gauge rosette listed in Table 6.1 is suitable. In such cases, it may be necessary to create a strain gauge rosette with dimensions or other features that are not included in the ranges supplied by manufacturers.

Where small gauges of a specific layout are required, an effective solution can be to order a "custom" strain gauge direct from a manufacturer. Figure 6.9 shows a small rosette with 031-size gauge elements based on the Type B layout (1/32"-size) with three elements located in a single quadrant. The gauge was developed to minimize the distances from the drilled hole to edges and holes on thin specimens. In this case, the gauge is of the open layout type to permit installation on (relatively) irregular surfaces. The proportions of some features of the rosette are different from those of existing gauges of this configuration because of the requirement for sufficiently large solder pads remote from the gauge elements.

Figure 6.9: Small, non-standard custom strain gauge rosette; installations adjacent to specimen features (photos courtesy of Stresscraft Ltd.).

Custom gauges of the type shown in Figure 6.9 can be manufactured to any reasonable scale for use with the appropriate drilled hole diameter and hole depth to satisfy the requirements of ASTM E837.

In cases where there is a requirement to determine residual stresses to depths greater than those associated with the standard gauges, it may not be practical or economic to produce custom rosettes of the large size needed. In these cases, strain gauge rosettes can be assembled from individual elements. The specimen surface is marked out with the hole center and gauge mean diameter lines (0.5D from the center). Individual gauge elements are aligned with the markings and bonded into position. In this way, successive installation of individual elements is a practical procedure that overcomes the difficulty of achieving a reliable, high-quality bond over a large area which would occur with an over-size rosette. However, great care must be taken to align the individual gauges accurately, else the measured strains will contain systematic errors.

Figure 6.10 shows an over-size rosette created from 1/4" gauge elements bonded to an aluminium specimen on the left side. The right side shows the rosette and specimen on completion of the drilling process.

By selecting gauge elements of the appropriate aspect ratio (Lg/Wg) ASTM E837 coefficients for Type A or Type B gauges can be used for the stress computation. For gauges with non-standard aspect ratios, stress computation coefficients are to be derived from finite element models. 1/4" gauge elements can be assembled in this way to detect residual stresses to a depth of 4 mm, while the maximum stress depth for rosettes assembled from gauge elements

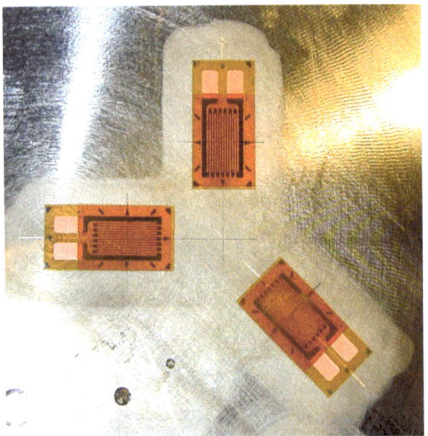

Figure 6.10: Large, non-standard 1/4" gauge rosette; installation and hole drilling (photos courtesy of Stresscraft Ltd.).

1/2" in length is 8 mm. Because of the comparatively large volumes of material to be removed, the drilling process for large gauges can be time consuming. Orbital drilling (circular milling) around the hole perimeter can reduce the machining time when compared with machining the entire hole. It can be demonstrated that any un-machined material remaining at the center of the hole has little effect on the strains detected by the gauge elements.

6.6   RESIDUAL STRESS EXAMPLE: TRAINING SAMPLE (ANNEALED DISC)

Figure 6.11 shows a set of residual stress distributions (c) obtained from strain gauge rosettes installed close to the bore of a titanium alloy disc (a) that had been machined and heat-treated. Image (b) shows one of three rosettes that were equi-spaced around the rear of the hub. The surface was prepared by fine abrasion using silicon carbide paper, following which the rosettes were bonded using cyanoacrylate adhesive. Rosettes were installed with element 1 aligned with the radial direction and element 3 in the circumferential direction. The residual stresses have been calculated from relaxed strains using the Integral Method, developed to correctly interpret residual stresses that vary with depth (Chapter 5).

The radial and circumferential stress distributions show modest levels of near-surface stresses, ~25 MPa, which correspond to less than 3% of the material yield strength. It can be seen that both radial and circumferential stresses increase with depth but do not exceed 40 MPa at the final increment depth (1024 μm). Generally, similar distributions were obtained from each of the three rosettes. The availability of a moderately stressed sample in a high strength material is a useful demonstration that the hole drilling process can be conducted in a reasonably "stress-free" manner. In particular, when the characteristics of a large sample of this type have been established, it can be used repeatedly for operator training, for quality testing drilling cutters and for developing hole drilling equipment and parameters.

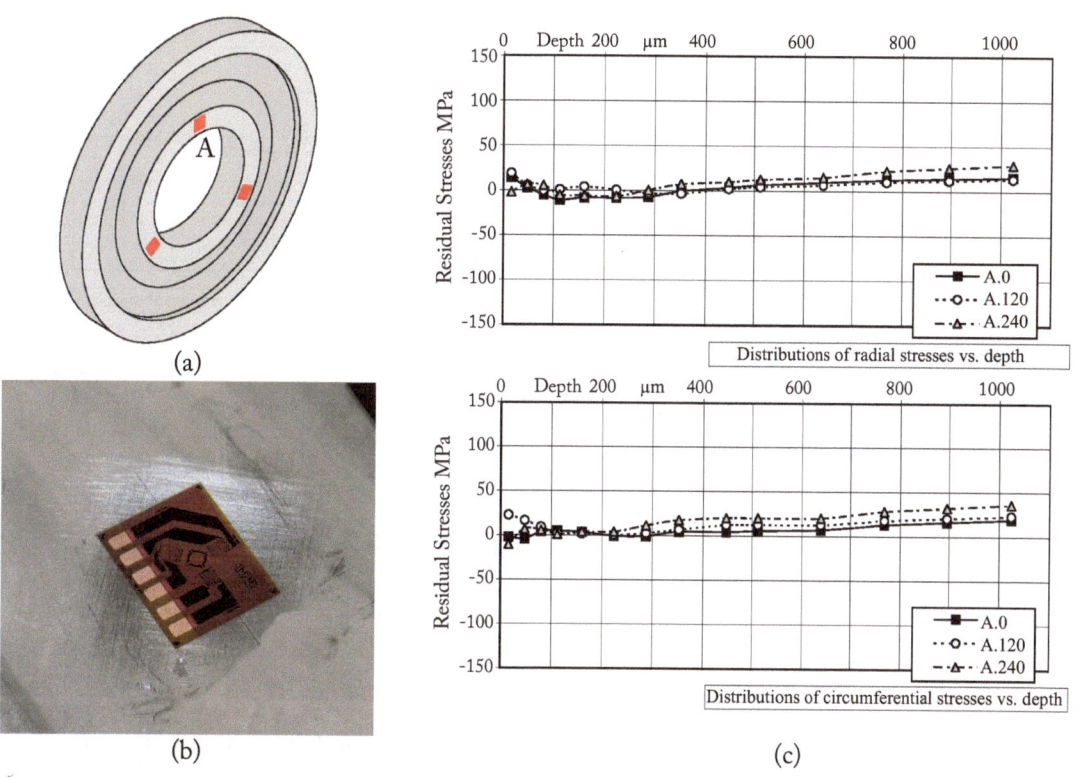

(a)

(b)

(c)

Figure 6.11: Annealed disc sample, gauge installation, and stress distributions (images courtesy of Stresscraft Ltd.).

6.7 RESIDUAL STRESS EXAMPLE: ALUMINIUM ALLOY BLOCK

The example shown in Figure 6.12 is a forged aluminium alloy block that has been subjected to various quenching and aging processes. The block has also been subjected to a deep etch to remove near-surface material containing residual stresses associated with machining processes. Two 1/8" open rosettes were installed and drilled at positions B and C to provide residual stress data to a depth of 2 mm. The distributions of stresses from the two rosettes shown here are typical of the results from a number of rosettes applied to this part. Because of the etching process, there are no significant disturbances in stress distributions close to the surface; the linear or uniform stress distributions extend from the sub-surface material to the surface.

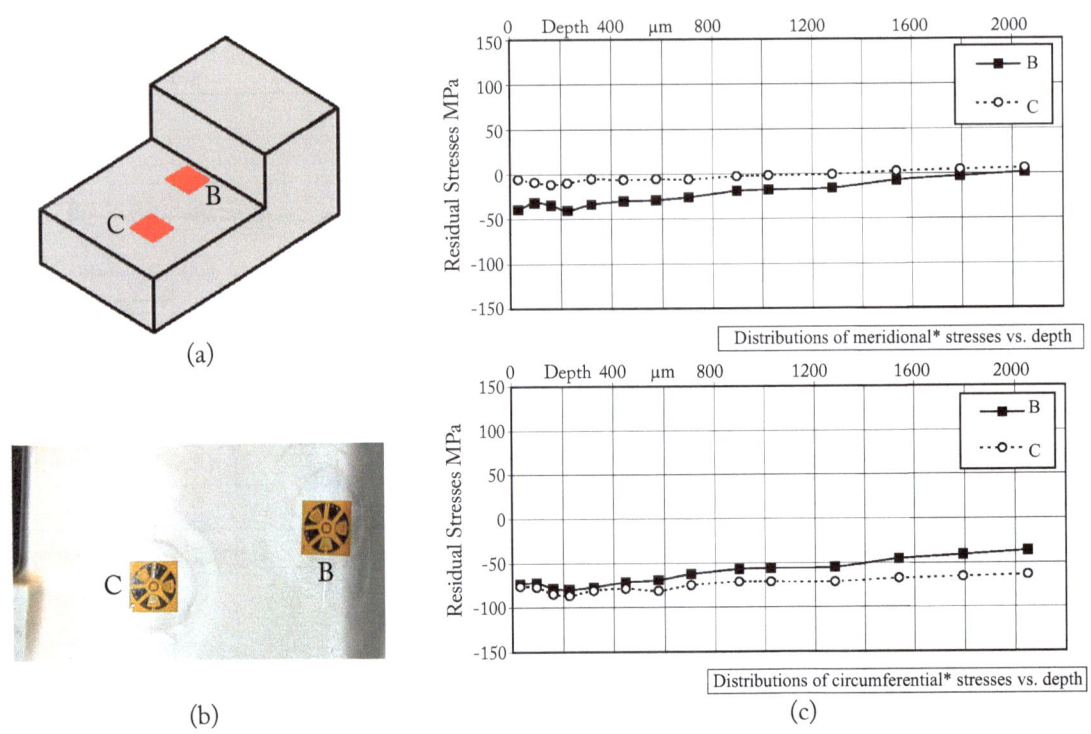

Figure 6.12: Aluminium alloy block sample, gauge installation, and stress distributions (images courtesy of Stresscraft Ltd.).

While hole drilling in this type of material presents no special challenge in terms of hardness, etc., the selection of drilling cutters and drilling parameters must be carefully controlled to avoid burring at the lip of the hole and the consequent disturbance to the near-surface stress field. It can be seen that this effect has been avoided in the stress distributions shown. As noted in the previous example, specimens of this type can be useful to demonstrate that the hole can be drilled without producing any drill-related artifacts. In this case, the linearity (or uniformity) of the stress distribution (or at any rate, the absence of significant discontinuities) can also provide useful confirmation of the performance of the stress calculation method, which is not addressed by the results from the low-stress sample.

6.8 RESIDUAL STRESS EXAMPLE: MACHINED, FORGED DISC

The disc nickel alloy disc shown in Figure 6.13a has been quenched and aged, resulting in significant levels of residual stresses throughout the disc structure before rough machining. At the side of the disc close to the bore, a 1/16" rosette (D) has been installed and drilled with "fine" increments (16 μm) close to the surface, becoming coarser at greater depths. Over the depth range 100–500 μm, the radial and circumferential stresses are distributed in a generally uniform manner, as shown in Figure 6.13c; nominal stress levels here are around −100 MPa (radial) and 300 MPa (circumferential). Closer to the surface, the stress input from the machining (lathe turning) process is clearly shown with tensile maxima in both radial and circumferential directions at the first calculation increment. The tensile stress in the direction of machining exceeds 400 MPa while the radial stress does not exceed 200 MPa at this depth. In this instance, the depth of material directly affected by the machining process appears to be within the first 100 μm.

While indicating the general shape of the stress distributions and levels of near surface stress gradients shown in Figure 6.13c, the Hole-Drilling Method with the equipment described here is not capable of determining the magnitudes of extreme stresses that exist very close to the surface in the first few microns. However, this result does provide a good illustration of how the Integral Method deals with large near-surface stress gradients. The stress distributions show that the near-surface stresses induced by the machining process are clearly tensile while sub-surface stresses at this part of the disc structure (caused by previous thermal processing) are compressive.

6.9 RESIDUAL STRESS EXAMPLE: SURFACE PROCESS SAMPLES

A series of titanium samples (50 mm × 50 mm × 10 mm thick) was prepared in which the conditions of the material surface were as follows.

E1. Wire-EDM cut surface

E2. EDM surface; recast layer removed using a fine abrasive

E3. Shot-peened

E4. Shot peened (as E3) and then smoothed to remove ∼50% of the peening indentations

E5. Shot-peened and smoothed (as E4) followed by further fine abrasion to remove the outer 50 μm of material.

Figure 6.14a shows a typical installation of a 1/32" pattern rosette on the surface of shot-peened sample 3. One gauge was installed on each of the samples and drilled using "fine" near-surface increments of 16 μm (becoming coarser with increasing depth). The resulting stress

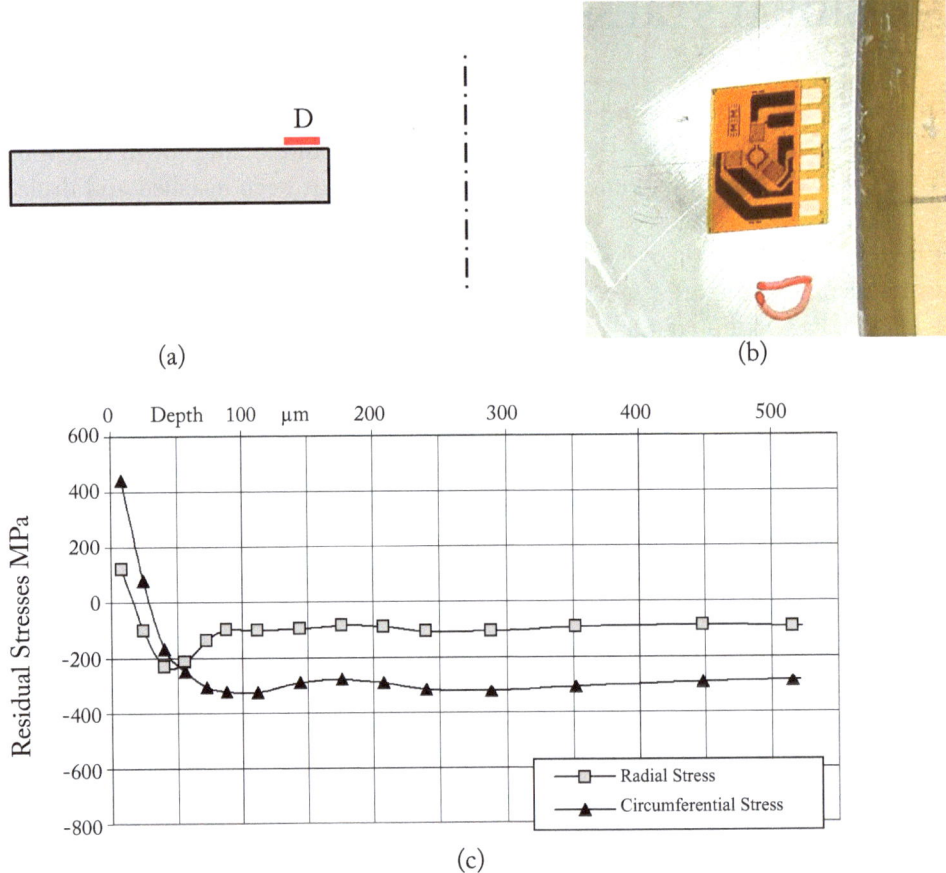

Figure 6.13: Machined, forged disc sample, gauge installation, and stress distributions (photos courtesy of Stresscraft Ltd.).

distributions (stress direction parallel to gauge element 1) are shown in Figure 6.14b. Features of the stress distributions include:

- low levels of sub-surface stresses in all samples; this helps the interpretation of the contributions of stresses made by the surface processes in this series;

- tensile stresses close to the surface of the wire-EDM cut re-cast layer (sample E1) in excess of 600 MPa; the tensile stress extends to a depth of ∼40 μm. Removal of this layer using fine abrasion and water coolant (sample E2), eliminates the near-surface tension;

- subsequent shot-peening of the sample results in compressive stresses exceeding 800 MPa over the depth range to 50 μm and extending to a depth of ∼150 μm (sample E3). Careful smoothing of the peened surface using water-cooled 400-grade silicon carbide paper results in little change in the stress distribution (sample 4); and

- after removal of a further 50 μm layer from the shot peened and smoothed surface (sample E5) using a honing process (water cooled), it can be seen that a significant proportion of the compressive layer remains; much of the near-surface distribution from sample 5 is similar in profile to samples E3 and E4, but shifted by an amount corresponding to the thickness of the material layer removed.

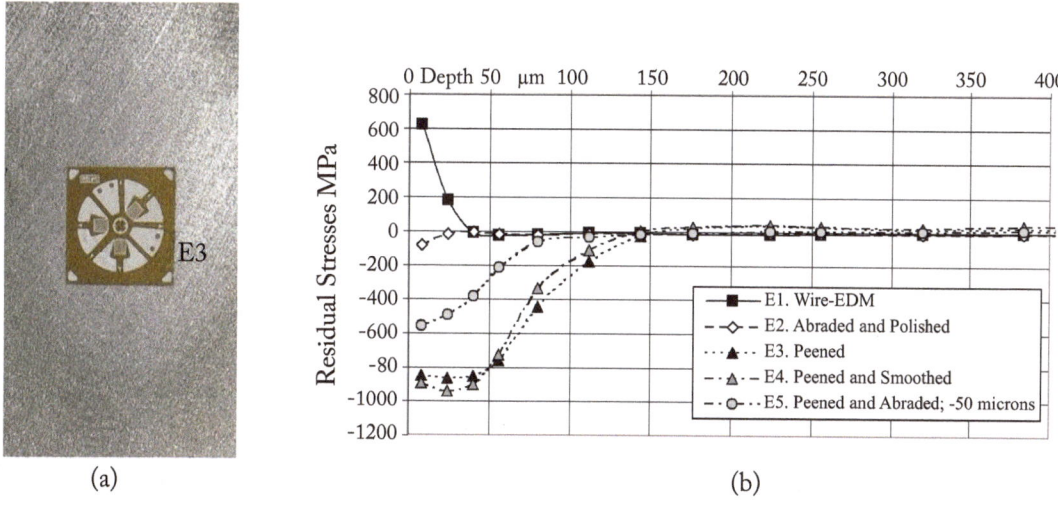

(a)   (b)

Figure 6.14: Surface process sample; gauge installation and stress distributions (photos courtesy of Stresscraft Ltd.).

6.10   RESIDUAL STRESS EXAMPLE: THIN, SHOT-PEENED BEAM

The example shown in Figures 6.15a and 6.15b is a steel alloy beam of cross section 0.9 mm × 10 mm that has been subjected to an intense shot peening process. Rosette F, a 1/32" Type A pattern, has been installed at the center of the top of the beam and drilled. A second rosette (G) has been installed on the underside of the beam, offset by 6 mm from the beam mid-length to reduce any interference from the previously drilled hole to an undetectable level. For each rosette, the hole drilling procedure has been carried out with the beam supported on a layer of cement. This prevents flexure caused by drilling forces to ensure control of the drill depth. The relaxed strain data from the two rosettes has been processed using the Integral Method incorporating coefficients derived from a series of finite element models of "intermediate" thickness plates.

Distributions of longitudinal stresses from the two gauges are shown in Figure 6.15c.

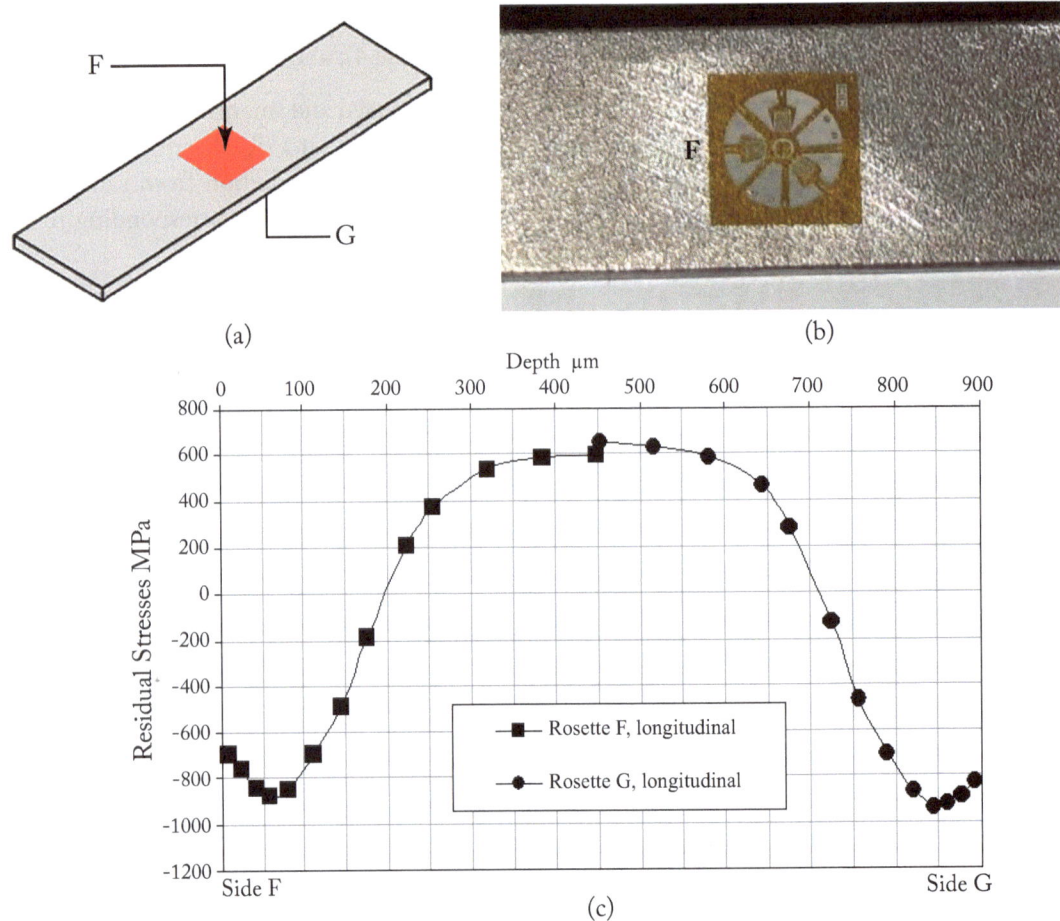

Figure 6.15: Shot peened steel beam, gauge installation and stress distributions (images courtesy of Stresscraft Ltd.).

- Intense levels of compression have been detected at the top and bottom surfaces with compressive stress maxima occurring at depths around 50 μm.

- Tensile stress maxima occur around the mid-thickness; the overall balance of compressive and tensile stresses for axial equilibrium appears to be satisfied.

- There is a reasonable level of symmetry in the combined stress distribution from the two gauges about the mid-thickness, with only a small dislocation in stress levels between the two distribution lines.

6.11 CONCLUDING REMARKS

For each of the examples described in the preceding sections, the practitioner is recommended to perform the necessary procedures of preparation, installation and hole drilling in a series of predetermined steps in line with the descriptions given in Chapter 4. While each of the individual steps may appear to be reasonably straightforward, the demands on the practitioner while performing the entire procedure can be severe. In order to produce data of the highest quality, it is recommended that the following provisions are made:

- adequate training in all aspects of the procedure and, in the case of infrequent practice, ample opportunity to rehearse the procedure on training specimens;

- appropriate working conditions, in terms of cleanliness, ambient temperature, noise levels and lighting;

- high quality and appropriately calibrated items of equipment and supplies of high quality consumable items; and

- sufficient time allowance to complete the work without interruption, and the opportunity to perform repeat measurements (at adjacent target sites), wherever required.

Wherever the practitioner detects shortcomings or defects (however small) in the procedures being followed or the equipment in use or perceives an opportunity to apply some improvement, then resources should be given to carry out experiments or trials so that such developments can be implemented to enhance the quality of future measurements. Every encouragement should be given to enhance the practitioner's skill to produce the best possible outcome.

6.12 FURTHER READING

- Koshti A, Egle DM (1985). An Alternate Technique for Implementing Center-hole Drilling/Residual-stress Measurements. *Experimental Techniques*, 9(12):28–30.

- Flaman MT, Herring JA (1986). SEM/ASTM Round-Robin Residual-stress-measurement Study-phase 1. *Experimental Techniques*, 10(5):23–25.

- Flaman MT, Herring JA (1986). Ultra-high-speed Center-hole Technique for Difficult Machining Materials. *Experimental Techniques*, 10(1):34–35.

- Plotkowski PD, Kowalski HC (1988). Residual Stress Determination on Nasty Surfaces. *Experimental Techniques*, 12(3):22–23.

- Lu J, Flavenot JF (1989). Applications of the Incremental Hole-Drilling Method for Measurement of Residual-stress Distribution. *Experimental Techniques*, 13(11):18–24.

- Chou GP, Lin YC (1993). Brief Approach for Aluminum Alloy Materials in Ultra-high-speed Hole-Drilling Technique. *Experimental Techniques*, 17(5):17–20.

- Whitney TJ, Stenger GJ (1993). A Device for Implementing the Strain Gage-hole Drilling Method of Residual Stress Measurement on Aircraft Transparencies. *Experimental Techniques*, 17(4):25–30.

- Nobre JP, Kornmeier M, Dias AM, Scholtes B (2000). Use of the Hole-Drilling Method for Measuring Residual Stresses in Highly Stressed Shot-peened Surfaces. *Experimental Mechanics*, 40(3):289–297.

- Valente T, Bartuli C, Sebastiani M, Loreto A (2005). Implementation and Development of the Incremental Hole Drilling Method for the Measurement of Residual Stress in Thermal Spray Coatings. *Journal of Thermal Spray Technology*, 14(4):462–470.

- Fontanari V, Frendo F, Bortolamedi T, Scardi P (2005). Comparison of the Hole-Drilling and X-ray Diffraction Methods for Measuring the Residual Stresses in Shot-peened Aluminium Alloys. *The Journal of Strain Analysis for Engineering Design*, 40(2):199–209.

- Sharman ARC, Hughes JI, Ridgway K (2006). An Analysis of the Residual Stresses Generated in Inconel 718™ When Turning. *Journal of Materials Processing Technology*, 173(3):359–367.

- Stefanescu D, Truman CE, Smith DJ, Whitehead PS (2006). Improvements in Residual Stress Measurement by the Incremental Center Hole Drilling Technique. *Experimental Mechanics*, 46(4):417–427.

- Rajendran R, Baksi P, Bhattacharya S., Basu S (2008). Evaluation of Non-uniform Residual Stress using Blind-hole Drilling Technique. *Experimental Techniques*, 32(3):58–61.

- Whitehead PS (2010). Practical Experiences in Hole Drilling Measurements of Residual Stresses. *Proc. of the SEM Annual Conference*, Society for Experimental Mechanics, 6(17):209-219, Indianapolis, IN, June 7–10.

- Valentini E, Beghini M, Bertini L, Santus C, Benedetti M (2011). Procedure to Perform a Validated Incremental Hole Drilling Measurement: Application to Shot Peening Residual Stresses. *Strain*, 47(s1):e605–e618.

- Scafidi M, Valentini E, Zuccarello B (2011). Error and Uncertainty Analysis of the Residual Stresses Computed by Using the Hole Drilling Method. *Strain*, 47(4):301–312.

- Nau A, Scholtes B (2013). Evaluation of the High-speed Drilling Technique for the Incremental Hole-Drilling Method. *Experimental Mechanics*, 53(4):531–542.

- Steinzig M, Upshaw D, Rasty J (2014). Influence of Drilling Parameters on the Accuracy of Hole-Drilling Residual Stress Measurements. *Experimental Mechanics*, 54(9):1537–1543.

- Ahmad B, Fitzpatrick ME (2017). Analysis of Residual Stresses in Laser-shock-peened and Shot-peened Marine Steel Welds. *Metallurgical and Materials Transactions A*, 48(2):759–770.

CHAPTER 7

Optical Techniques

"The development of numerous optical approaches for use with hole drilling may be considered evidence of the technological significance of the hole drilling method as a means for determining residual stresses."

Drew Nelson (2010) in "Residual Stress Determination by Hole Drilling Combined with Optical Methods."

7.1 INTRODUCTION

The history of the Hole-Drilling Method has been marked by a continuous development of all aspects of the technique. Great advances have been made in equipment, measurement methods and in computational procedures. The introduction of strain gauges in the late 1940s marked a turning point in the evolution of the Hole-Drilling Method and enabled it to grow into the accurate and reliable residual stress measurement technique that it is today. However, that is not to say that strain gauges are the only suitable way to make hole-drilling measurement, nor that they are always the ideal choice. Like all other techniques, strain gauges have their particular strengths and also some concerns.

Since the 1980s, optical techniques have been explored as an alternative means of measuring the surface deformations that are created during the hole-drilling process. These techniques have the advantage of providing full-field data, which are useful for data averaging, error checking and extraction of detailed information. Effectively, having full-field optical data is like having multi-element strain gauge rosettes of the type shown in Figure 5.1, but with numerous thousands of available gauges. In many ways, the optical techniques are complementary to the strain gauge technique, each approach having generally opposite advantages and disadvantages. Table 7.1 lists some features of strain gauge and optical measurements.

Two general classes of optical methods have been applied to hole-drilling; Interferometry and Digital Image Correlation (DIC). These methods have their origins as general metrology techniques and each has a large and active research literature. They have been adapted to hole-drilling residual stress measurements, for which a specialized literature is also rapidly growing. The following sections describe some of the several optical approaches explored.

Table 7.1: Features of strain gauge and optical measurements

Strain Gauge Measurements	Optical Measurements
Moderate equipment cost, high per-measurement cost	Higher cost equipment/software, moderate per-measurement cost
Significant preparation and measurement time	Preparation and measurement time can be short
Small number of very accurate and reliable measurements	Large number of moderately accurate measurements that can be averaged
Stress calculations are relatively compact	Stress calculations often quite large
Modest capabilities for data averaging and self-consistency checking	Extensive capabilities for data averaging and self-consistency checking
Relatively rugged, suitable for field use	Less rugged, more suited to lab use
Sensitive to hole-eccentricity errors	Hole center can be identified accurately

7.2   HOLOGRAPHIC INTERFEROMETRY

The concept of holography was introduced by Dennis Gabor in 1947, for which development he was recognized by the Nobel Prize in 1971. The process involves illuminating an object by coherent monochromatic light in an interferometer. Light scattered by the object forms an interference pattern when mixed with reference light from the same light source. The pattern can be recorded on very fine grain photographic plate. The fine grain of the photographic plate allows the interference pattern to be recorded on a wavelength scale such that both phase and amplitude data are recorded. When the resulting photographic image is subsequently illuminated by a reference light, a virtual image of the original object can be observed. This is the basis for the process used to create the holographic images that are in common use in credit cards, security seals and novelty items.

In the classical technique, the "recording medium" is a photographic plate. This works effectively, but is inconvenient experimentally because of the time and effort required to develop the plate. A practical alternative approach is to use a thermoplastic hologram plate instead. This plate does not require chemical treatment and is "developed" in place by electrostatic means. In addition, after the contained fringe pattern has been recorded, the plate can easily be erased and reused for subsequent measurements.

Figure 7.1 illustrates the holographic technique used for hole-drilling measurements. It is a variant method that makes a hologram by exposing a thermoplastic recording plate to reference and scattered object light and then developing the plate. After hole-drilling the plate is exposed again to the same light sources. The hologram then displays a pattern of live light and dark "fringes" that resemble a contour map. Figure 7.2 illustrates an example holographic fringe pattern. The fringes indicate lines of constant surface displacement between holographic expo-

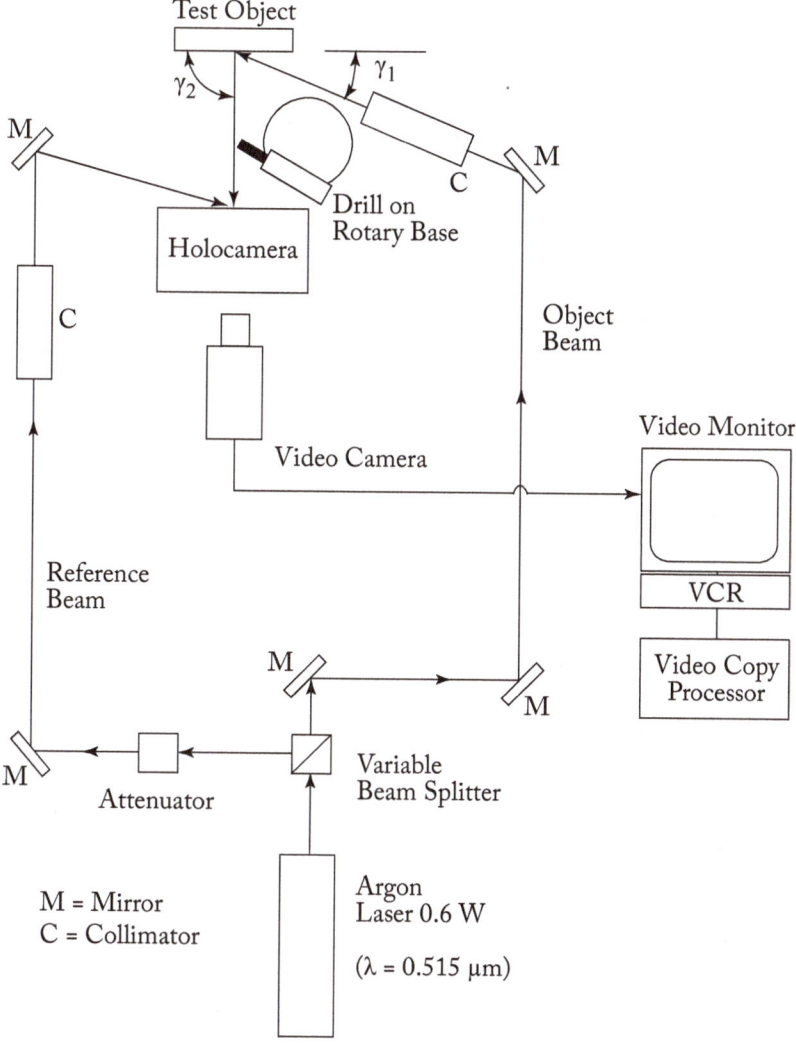

Figure 7.1: Example optical arrangement for making holographic hole-drilling measurements (from Nelson et al. (1994)).

sures. The displacement interval between fringes is a fixed value that depends on the wavelength of the illumination light and the geometry of the illumination arrangement used. The example pattern is unsymmetric because the sensitivity direction of the measurement is inclined to the right, midway in direction between the incident and reflected beams at the test object surface.

The apparatus in Figure 7.1 consists of a laser source to provide the required coherent monochromatic illumination. A beam splitter divides the output into two separate beams, the

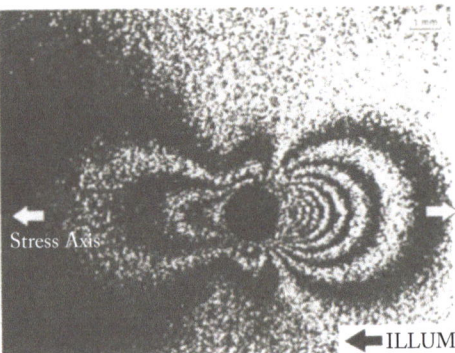

Figure 7.2: Example holographic fringe pattern (from Nelson et al. (1994)).

reference beam going directly to the recording medium ("holographic camera"), and the object beam first illuminating the test object surface and then reflecting onto the recording medium. The two beams mix there and form an interference pattern which is stored as a hologram by developing the recording medium. After the initial interference pattern exposure, a drill on a rotary base is used to cut a hole in the specimen and the recording medium is then re-exposed to reference and object beam light. A pattern of dark and light interference fringes appears and is viewed through a video camera.

A characteristic of the holographic method is that the fringe patterns are observed in analog format as a sequence of light and dark lines. Amplitude information is available but not sign, i.e., it is not known whether the fringes represent tension or compression. However, the sign of the associated deformations can be determined by a subsequent measurement where deformations of known sign are deliberately applied to the test object and then noting whether the observed fringes move closer together or further apart. Alternatively, the same effect can be achieved by slightly rotating the incident beam on the test object. This superposes an artificial tensile or compressive field over the observed fringe pattern, depending on the beam rotation direction. In the latter case, the superposed fringes are called "carrier fringes."

7.3   MOIRÉ INTERFEROMETRY

Moiré interferometry is a variant interferometric technique that has been investigated for hole-drilling residual stress measurements. Figure 7.3 schematically shows a typical optical arrangement. The interferometer arrangement is similar to that in Figure 7.1, but has some significant differences. Light from a single coherent laser source is split into two symmetric beams that illuminate the specimen surface. A diffraction grating consisting of finely ruled lines, typically 600–1200 lines/mm, is replicated or made directly on the specimen surface. Diffraction of the

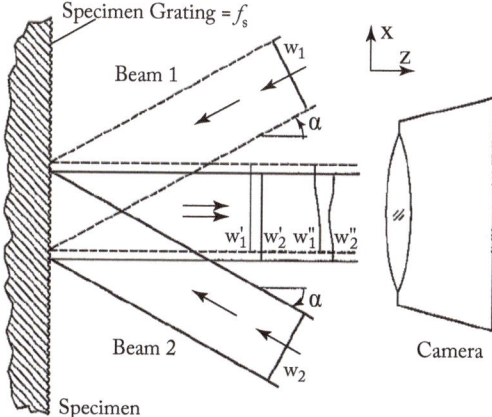

Figure 7.3: Example optical arrangement for making Moiré hole-drilling measurements (from Wu et al. (1998)).

light beams creates a diffraction pattern, giving interference fringes consisting of light and dark lines, similar in appearance to those observed with the holographic technique.

Figure 7.4 illustrates some typical measured Moiré fringe patterns. As before, these fringe patterns represent contour lines of constant surface displacement. For the symmetrical beam arrangement in Figure 7.4, the measurement sensitivity direction is in-plane in the x-direction. The fringes from the Moiré method have the same analog character as those from the holographic method and also do not directly indicate sign. This indeterminacy can again be resolved

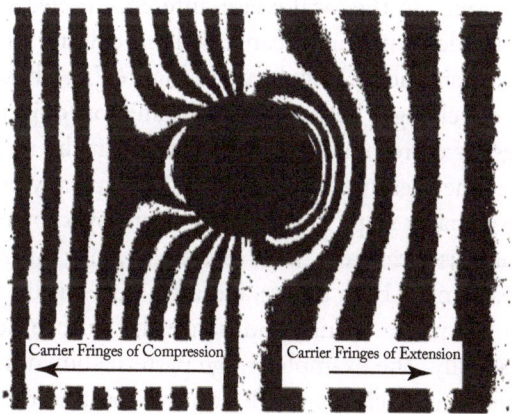

Figure 7.4: Example Moiré fringe patterns with supposed tensile and compressive carrier fringes (from Wu et al. (1998)).

by the slight rotation of one of the beams or specimen to produce carrier fringes that super-impose artificial tensile or compressive strains across the measured fringe patterns. Figure 7.4 illustrates some Moiré fringe patterns measured with superposed tensile and compressive carrier fringes.

Both the Moiré and holographic techniques have the advantage of making useful measurements very near to the hole boundary, much nearer than could be made by strain gauges. When using an attached diffraction grating with the Moiré method, some minor delamination of the grating near the hole edge can occur, which will limit the closeness of available measurements. The surface preparation to attach or form the diffraction grating on the specimen surface is a drawback of the Moiré method, but is not prohibitive.

7.4   ELECTRONIC SPECKLE PATTERN INTERFEROMETRY (ESPI)

Electronic Speckle Pattern Interferometry (ESPI) is a popular interferometric measurement technique that is widely used for surface metrology. It uses a video camera to record interference pattern images. It is an attractive measurement approach because it avoids the need with holographic interferometry for a thermoplastic plate or similar recording device.

Figure 7.5 shows a typical ESPI arrangement. The light from a coherent laser source is divided into two beams, one of which illuminates the specimen surface and is subsequently imaged by a digital camera, while the second feeds directly to the camera where it creates an interference pattern on the sensor surface. The interference pattern, such as shown in Figure 7.6a, appears as random speckles spread across the measured image. At first glance the interference pattern appears as if it contains just noise. This is partly true inasmuch as there is a random phase variation from one pixel to the next. However, each pixel considered alone behaves systematically and has a phase that corresponds to the local phase difference between the illumination/object and reference beams. Surface deformations created by hole-drilling cause small changes in the optical path length of the illumination/object beam in Figure 7.5. These path length changes in turn cause proportional phase changes between the illumination/object beam and the reference beams, which then manifest as corresponding phase changes at the measured image pixels.

The phase changes at the measured image pixels are typically evaluated using the phase stepping method. A piezo actuator steps the reference beam by known phase angle steps, after each of which a speckle image is measured. Most commonly, four images are taken at 90° intervals. These measurements are made both before and after hole-drilling, producing a combined set of images from which the phase change and hence the surface deformation at each pixel can be evaluated mathematically. The example fringe pattern shown in Figure 7.6b graphically represents the phase evaluated at each point. These results define the phase within the cyclic range $-\pi < \phi \leq \pi$ and are called "wrapped." A further process called "unwrapping" is required to arrange the phase data in continuous format so that they then properly represent the surface displacements.

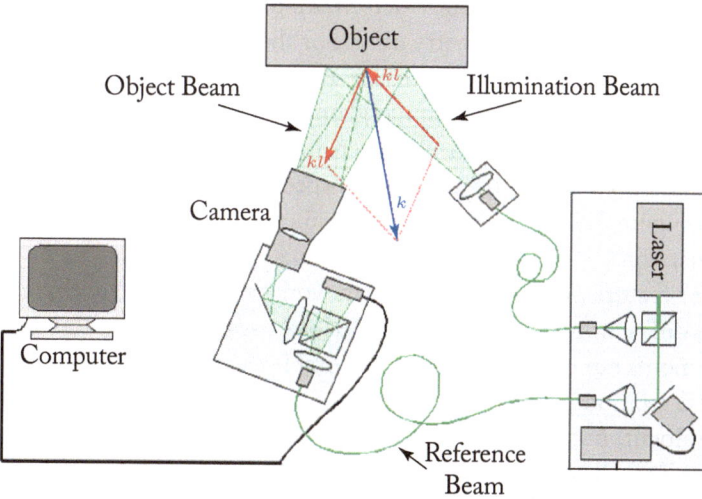

Figure 7.5: Schematic diagram of a typical ESPI apparatus used for hole-drilling measurements (adapted from Steinzig and Ponslet (2003)).

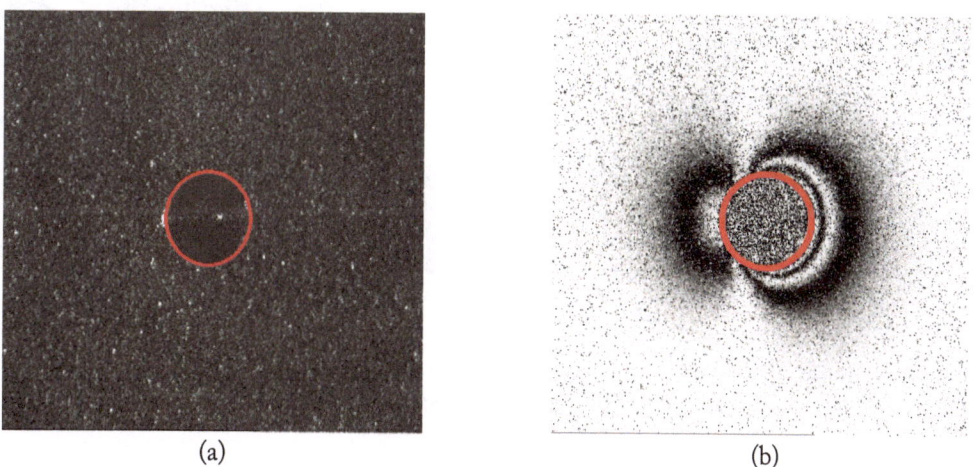

Figure 7.6: ESPI hole-drilling measurements. (a) Speckle image and (b) fringe image.

The phase-stepping technique is particularly attractive because it produces a numerical value of the surface deformation at each pixel location, including sign. This is big advantage over the analog unsigned data from photographic holography and allows much more detailed and sophisticated data analysis. The phase-stepping technique can also be used with the Moiré method illustrated in Figure 7.3 and provides similar advantages.

The relationship between the surface deformations and the resulting ESPI phase changes depends on the particular optical geometry used. For the arrangement shown in Figure 7.5, the sensitivity direction lies along the blue vector "k" that bisects the directions of the illumination and object beams. An ESPI setup with similar geometry to that used for the Moiré interferometer shown in Figure 7.3 would have in-plane sensitivity. When used for ESPI measurements, a diffraction grating would not be needed on the specimen surface. Both the out-of-plane geometry shown in Figure 7.5 and the in-plane geometry shown in Figure 7.3 are suitable for hole-drilling applications.

A significant feature of ESPI is that it can work with a plain specimen surface, without attachment of the diffraction grating needed for Moiré measurements. This makes it possible to do ESPI measurements rapidly, thus making it suitable as an industrial quality control tool. As with all the optical methods, ESPI equipment is more complex and expensive compared with strain gauge equipment, but the per-measurement cost can be significantly lower. Figure 7.7 shows examples of ESPI devices suitable for industrial use.

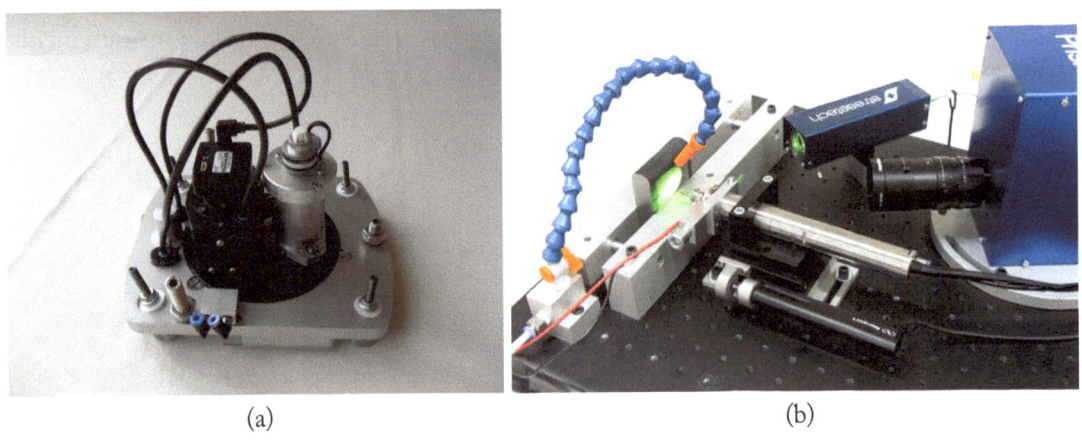

(a)   (b)

Figure 7.7: Example ESPI equipment suitable for industrial use. (a) Radial interferometer (A. Albertazzi and M. Viotti) and (b) PRISM system (Stresstech Oy).

7.5   DIGITAL IMAGE CORRELATION

Digital Image Correlation (DIC) is a versatile general-purpose optical technique used for measuring surface displacements in two or three dimensions. Figure 7.8a schematically shows a typical arrangement used for 2-D (in-plane) displacement measurements. The experimental procedure involves painting a textured pattern on the specimen surface and imaging the region of interest using a high-resolution digital camera. The camera, which is set perpendicular to the surface, records images of the textured surface before and after deformation. The local details within the two images are then mathematically correlated, and their relative displacements determined. The algorithms used for doing this have become quite sophisticated, and with a well-calibrated optical system, displacements of $+/-0.02$ pixel can be resolved.

The 3-D technique, schematically illustrated in Figure 7.8b, involves imaging the region of interest with two cameras and using stereoscopic imaging to determine deformations in three dimensions. The equipment is more complex than for the 2-D technique, and careful setup and calibration are required. For hole-drilling measurements the out-of-plane displacements are small compared with the in-plane displacements, so in this case the extra data available from 3-D measurements provide only modest advantage.

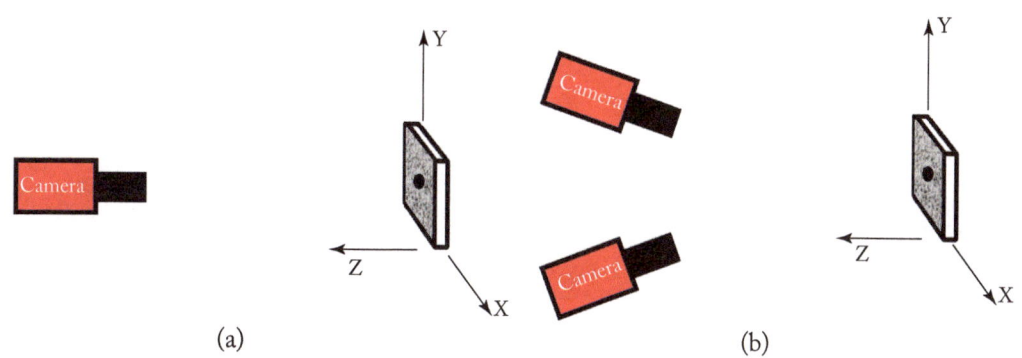

Figure 7.8: Typical DIC measurement arrangements, (a) 2-D and (b) 3-D.

The DIC method has both advantages and disadvantages compared with the interferometric methods. Its major disadvantage is its relatively low sensitivity, about an order of magnitude less when measuring a typical ~2 mm diameter hole. It also requires the painting of a speckle pattern on the specimen surface. However, once the pattern has been applied, the surface is quite durable and can be cleaned if it becomes dirty. By comparison, a surface for ESPI measurements cannot be touched at all; even very minor abrasions will damage measurement quality. In addition, DIC can directly evaluate displacements in two in-plane directions, also the out-of-plane direction if the 3-D technique is used. The large amount of available data improves the quality of the resulting residual stress evaluation. By comparison, typical ESPI systems provide displacement data in only one direction. Multi-axis ESPI systems do exist, but are uncommon.

DIC measurements differ in a very significant conceptual way from interferometric measurements. Interferometric measurements are size dependent, with spatial sensitivity defined relative to the wavelength of the light source used. In contrast, DIC is size independent because its sensitivity depends only on the pixel density of the imaged area, not on the physical dimensions represented within the image. A drilled hole could be large or small, but its appearance within a measured image is the same. Thus, the DIC analysis procedure is the same. Consequently, DIC is an attractive choice for hole-drilling measurements where the hole size is either much larger or much smaller than the 0.5–5.0 mm diameter range suited to interferometric measurements. Figure 7.9 shows examples of such measurements, part (a) showing a 63.5 mm diameter hole and part (b) showing a 1.5 μm diameter hole observed within a scanning electron microscope.

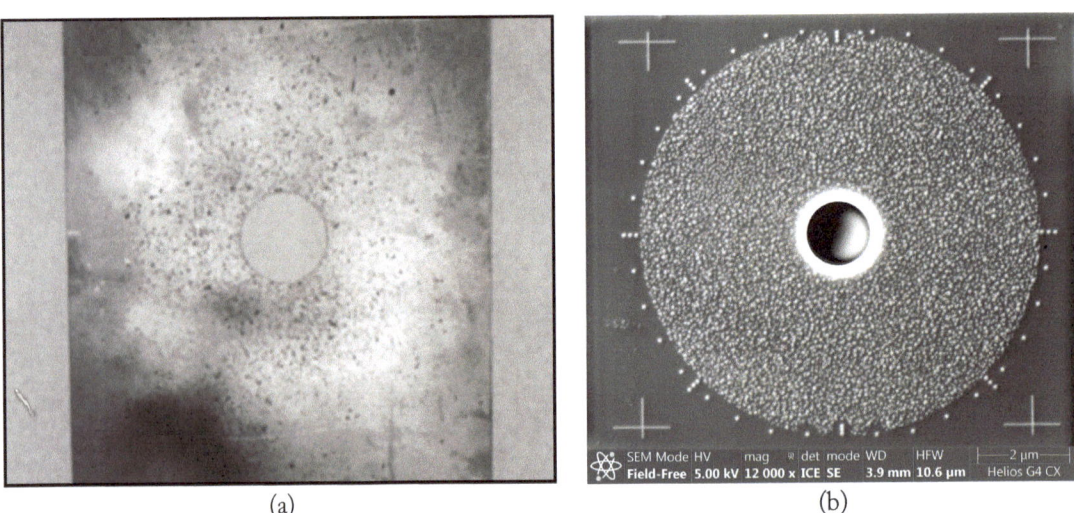

(a)   (b)

Figure 7.9: DIC hole-drilling residual stress measurements at different scales. (a) hole diameter = 63.5 mm (photo from McGinnis et al. (2005)) and (b) hole diameter = 1.5 μm (image courtesy of B. Winiarski).

7.6   COMPUTATION OF UNIFORM RESIDUAL STRESSES

The availability of full-field data from optical measurements presents both opportunities and challenges. The rich data set provides many possibilities for productive data analysis, for example, to do data consistency checking and outlier correction, data averaging, evaluation of additional stress results and measurement artifact compensation. The large size of the data set also creates challenges as how to handle the substantial bulk of the data in an effective way. Some desired objectives of a residual stress computation method for use with optical data include to:

- take advantage of the wealth of data available within an optical image;

- extract the data from the image with a minimum of human interaction preferably none; and

- use the available data in its native form in a compact and stable computation.

Initial optical measurements for hole-drilling residual stress evaluations used calculation methods analogous to those for use with strain gauges. Typically, they involved visually picking a small number of opportune points within the measured images, interpreting their fringe orders, and then doing a strain gauge style calculation. Such fringe pattern interpretation is an important need when working with the unsigned analog data that comes from traditional holographic measurements with photographic or thermoplastic plates. Although effective results are achieved, the performance of these methods can be further enhanced by including the contributions of the substantial quantity of additional data available beyond the few selected points.

The ESPI, phase-stepped Moiré and DIC methods share the advantage that they provide signed numerical data. A typical image taken with a modern video camera contains several hundred thousands, often millions of pixels, thus a very large number of independent measurements are produced. The availability of such large amounts of quantitative data makes these methods well suited for more in-depth data analysis.

In contrast to the traditional strain gauge style measurements, all optical techniques indicate surface displacements, not strain. Estimation of surface strains from displacement measurements involves numerical differentiation, which is an inherently noisy process, and consequently is to be avoided. Thus, it is desirable to choose computation methods that work directly with displacement data. In addition, linear methods are particularly desirable to achieve effective and compact data processing. Nonlinear procedures can also be effective, but they are much more computationally intensive, potentially less stable, and thus should be used only when essential.

A computation that has a greater number of data than unknowns is called "overdetermined." The number of available optical data is much greater than the number of unknowns, so there exists a good opportunity to extract some further results from the measurements. Here, it is very effective also to include in the calculation the effects of systematic artifacts that can occur in practical experimental measurements. For example, in each of the various optical methods, drifts often occur that cause the data from all image pixels to shift by the same amount. In the ESPI and Moiré methods such drifts occur from temperature changes within the apparatus, with DIC they occur from small position changes of the optical components. Similarly, image stretching can occur due to specimen temperature changes, or to magnification changes in the case of DIC measurements within a scanning electron microscope. Even with these additional factors included in the calculation, the residual stress computation is still highly overdetermined. The excess data can usefully be exploited to provide an averaging effect that acts to reduce the effect of random measurement noise.

It is generally impossible to achieve a numerical solution of an overdetermined measurement that exactly fits all measured data. Consequently, a "best-fit" solution is sought that is as close as possible to the majority of the measured data. This can conveniently be done using the Least-Squares method. This approach was introduced for ESPI measurements by Steinzig and Ponslet (2003) and further developed by subsequent authors. Figure 7.10 schematically describes the procedure. The starting point is the measured image shown on the left. This could be an unwrapped phase map measured using the ESPI or Moiré methods, or a displacement map from the DIC method. On the right side are the computed responses to the factors that are anticipated to contribute to the measurement. In this case, these are the P, Q, and T stresses and three potential data artifacts. In this single-axis example, where the surface displacements have been measured in the x-direction only, the artifacts, respectively, are bulk displacement, uniform stretch and uniform shear, all in the x-direction. For dual axis measurements, the analogous artifacts in the y-direction could further be included.

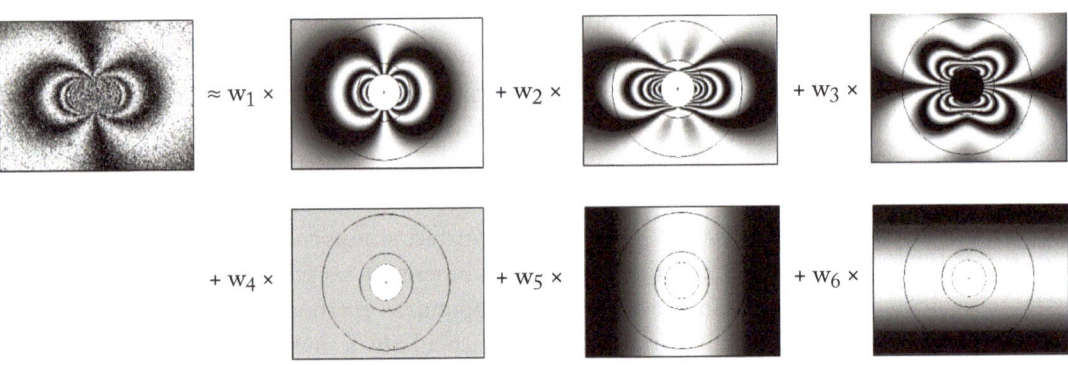

Figure 7.10: Least-squares fit of a measured fringe pattern as a weighted sum of six contributing factors. $1 = P$-stress, $2 = Q$-stress, $3 = T$-stress, $4 = x$-displacement, $5 = x$-stretch, and $6 = x$-shear.

The six contributing factor images shown in Figure 7.10 represent their individual responses when each of them has a unit size. The actual sizes of the contributing factors are represented by the weights w_i, $1 \leq i \leq 6$ that appear as numerical factors. The measured response on the left can then be interpreted as the weighted sum of the responses on the right, with each of the right-side source quantities present in the corresponding amounts in the measured specimen. The weighting factors w_i can be determined using the Least-Squares method. It finds the combination of the weighting factors that gives a sum that most closely fits the measurements on the left of Figure 7.10.

Figure 7.11 shows the "best-fit" sum of the six contributing factors included in Figure 7.10. Because of the overdetermination of the calculation, the fit is not exact at every point. The remaining difference between the measured data and the sum of the best-fit data is called the "residual." Ideally, the residual should appear as in Figure 7.11, with uniformly random noise and without apparent structure.

Practical calculations proceed with the data contained in the images in Figure 7.10 arranged in vector-matrix format

$$Gw = \delta, \qquad (7.1)$$

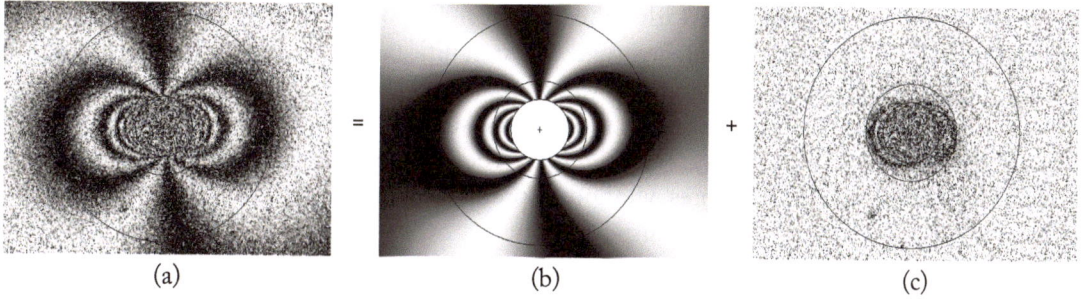

Figure 7.11: Least-squares fit fringe pattern. (a) Measured fringe pattern, (b) best-fit fringe pattern, and (c) residual.

where the vector w contains the six weighting factors w_i. Vector δ contains the displacements represented by the N pixels in the measured image that are used for the calculation. These pixels are taken from the annular area around the drilled hole shown in the image on the left of Figure 7.10. This area is chosen to encompass a region that:

- is circular so as to minimize directional bias in the calculation;

- has an inner radius sufficient to exclude pixels damaged by chips scratching the surface as they exit the drilled hole. However, it should not be so large so as to lose valuable high-displacement data near the hole edge. In the example in Figure 7.10, the inner radius was chosen as 1.8 times the hole radius, which excludes most of the damaged pixels visible in the residual plot, Figure 7.11c; and

- not too distant from the hole so as only to include pixels with significant displacement data content. Pixels far away from the hole are excluded because the potential benefit of their displacement data content becomes small compared with the potential damage from their noise content.

The choice of the inner and outer radii of the selected annular area requires some judgment. Typical values of the inner radius are between 1.5 and 2.0 times the hole radius. Typical values of the outer radius are between 3.0 and 4.0 times the hole radius. The latter is not a sensitive choice and can be selected to fit the scale of the measured image. To facilitate a reasonable choice, the magnification used for the initial setup of the optical arrangement for a hole drilling measurement should be set to make the height of the measured image approximately four to five times the anticipated hole diameter.

Matrix G has N rows corresponding to the pixels in δ and 6 columns corresponding to the unknowns in w. The contents of the columns are the six unit contributing factors shown on the right of Figure 7.10, arranged in the same pixel order as in vector δ. Since the three artifacts represent physically different things from the three stresses, it can happen that the corresponding columns of G have elements of greatly differing sizes. This occurrence would seriously impair the numerical conditioning of the matrix and create significant roundoff error effects. This issue can be avoided by choosing units for the artifact quantities that give columns in matrix G that have contents of generally similar size to those in the stress columns.

Since $N >> 6$, Equation (7.1) is highly overdetermined. The "best-fit" solution can be determined using the Least-Squares Method by premultiplying the equation by G^T, where the superscript T indicates the matrix transpose

$$G^T G w = G^T \delta. \tag{7.2}$$

By paying attention to the sequence of the required multiplications, it is possible to form the 6×6 $G^T G$ matrix and the 1×6 right-side vector $G^T \delta$ directly by accumulating the various permutations of the dot products of the matrix columns and displacement vector. This procedure minimizes the required numerical effort by avoiding the explicit creation and handling of the very large matrix G and right-side vector δ. Additional computational efficiency can be achieved by noting that matrix $G^T G$ is symmetrical, so only half of it need be computed explicitly. In addition, several of its contents are known in advance to be zero because they represent the dot products of even and odd functions, for example w_1 and w_3. The resulting order-6 matrix equation can be solved routinely. Substitution of the resulting w values back into Equation (7.1) gives the corresponding best-fit displacement response shown in Figure 7.11b. The residual in Figure 7.11c shows the difference between this and the actually measured displacement response in Figure 7.11a. As hoped, the residual is random, without significant structure. This confirms that the six chosen contributing factors have successfully modeled most of the observed response.

A further computational opportunity exists when working with DIC data or with data from a 2-axis ESPI setup. Both of these arrangements provide separate displacement data in the x and the y directions, one each at each measured pixel. In the case of DIC, the x and y displacements are computed together at each pixel, and in the case of 2-axis ESPI, the data derive from separate x and y displacement images. Analysis of these measurements to determine residual stresses is a direct extension of the above procedure, with the structures of the vector-matrix quantities in Equations (7.1) and (7.2) expanded to include the second axis data. In addition, three further contributing factors are needed to describe displacement, stretch and shear artifacts in the y direction. The total of contributing factors then becomes 9, comprising 3 stresses plus 3 x-artifacts plus 3 y-artifacts. Figure 7.12 schematically illustrates the nine items.

Figure 7.13 schematically illustrates the structure of Equation (7.1) for the 2-axis case. For simplicity, the diagram shows the first six pixels to represent the much larger group, with non-zero matrix coefficients indicated by asterisks. Here, the solution vector w has expanded to length 9 and the displacement data vector δ has doubled in length to $2N$, where N is the number of pixels used in the calculation. The first half of the vector contains the measured x-displacements and the second half the y-displacements. Correspondingly, matrix G has expanded to have nominal size $2N \times 9$.

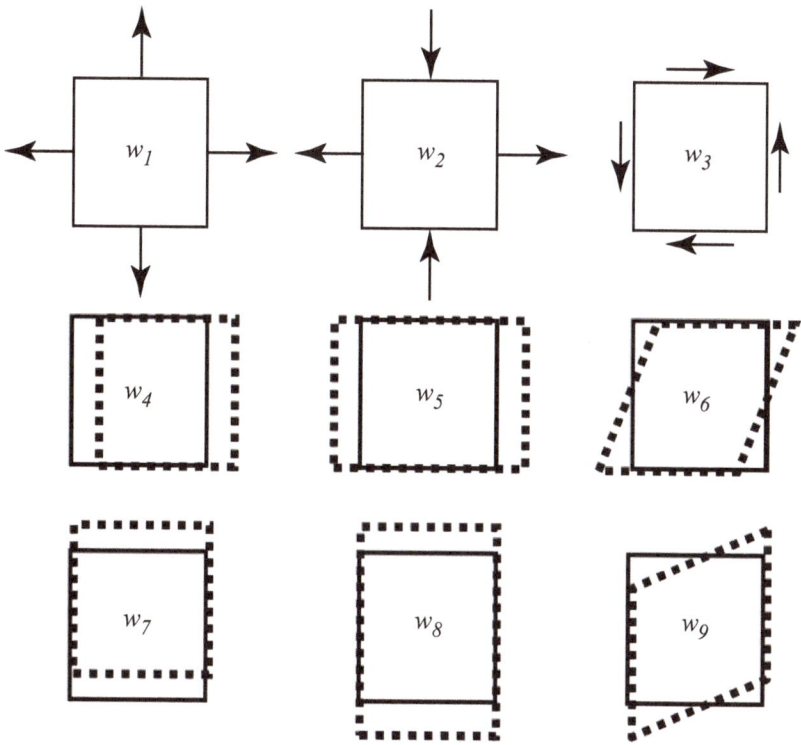

Figure 7.12: Schematic representation of the contributing factors to a 2-axis hole-drilling measurement. $1 = P$-stress, $2 = Q$-stress, $3 = T$-stress, $4 = x$-displacement, $5 = x$-stretch, $6 = x$-shear, $7 = y$-displacement, $8 = y$-stretch, and $9 = y$-shear.

As can be seen, G has a sparse structure because the artifacts associated with w_4, w_5 and w_6 are connected only with the x-displacements and the artifacts associated with w_7, w_8 and w_9 are connected only with the y-displacements. Computational efficiency can be enhanced by taking the sparsity structure of the 2-D matrix G into account so as to avoid handling the zero elements. In addition, the elements in the lower half of the first three columns of G are the same as those in the upper half, with the x and y coordinates interchanged. This corresponds to reflecting the first three images in Figure 7.10 around a 45° diagonal. Similarly, the contents of the lower parts of the second three columns are the same as those in the upper parts of the third three columns, again the x and y coordinates interchanged. Actually, providing the concept of exchange of x and y coordinates is followed, the pixels in the upper half of G do not have to be exactly those in the lower half, they don't even need to be equal in total number. This feature occurs because the solution depends on the combined behavior of many pixels, not the individual behavior of specific pixels. This is a significant point when working with ESPI data, where the second axis data comes from a different image than the first.

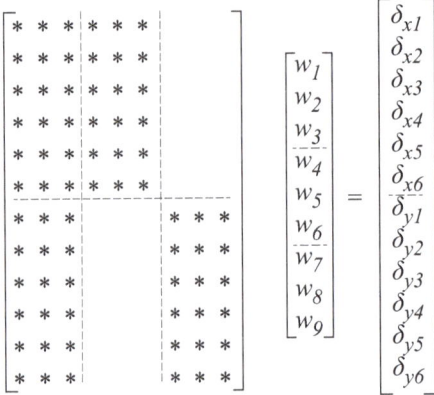

Figure 7.13: Schematic diagram of the structure of Equation (7.1) for a 2-axis measurement.

With care about the handling of the numerical data, solution of Equation (7.2) for the 2-axis case is not very much more burdensome than the single-axis case, just over twice the total computation time. The extra measurement and computation effort brings two significant benefits.

1. **Isotropic measurement sensitivity.** Residual stresses in all directions are evaluated with the same precision. In contrast, evaluation of the residual stress in the perpendicular direction of a single-axis measurement is governed by Poisson's ratio and its precision is reduced by that factor.

2. **Greater expected accuracy.** A two-axis measurement uses $2N$ data to determine nine quantities, while a single-axis measurement uses N data to determine six quantities. A 2-axis measurement can therefore be expected to produce a more accurate result because it has available one and a half times the data per evaluated quantity.

7.7   COMPUTATION OF NON-UNIFORM RESIDUAL STRESSES

The calculation procedure described in the previous section is effective to determine hole-drilling residual stresses for the case where the stresses are uniform within the hole depth. It is the full-field optical equivalent of doing a strain gauge uniform stress calculation of the type described in Chapter 5, Section 5.2. However, it most commonly happens that the residual stresses in a material are not uniform with depth from the specimen surface. The approach required to evaluate the variation of stress with depth from full-field optical data parallels that required for strain gauge data, described in Section 5.6. In summary, the required procedure is as follows.

1. Drill a hole in a series of small depth increments down to the final desired depth.

2. Make an optical measurement of the surface around the hole after each hole depth increment has been completed.

3. Discretize the residual stress profile into a series of steps based on the hole depth increments (see Figure 5.7).

4. Use the Integral Method to determine those stresses from the combination of all the measured data.

In concept, the procedure for implementing the Integral Method with full-field optical data substitutes Equation (7.1) for the analogous strain gauge Equations (5.18). Figure 5.8 in Chapter 5 schematically illustrates the computational concept of the Integral Method. The diagrams represent the responses of the material around a drilled hole to the residual stresses within various hole depth increments in holes of various depths. The measured response at a given hole depth is the sum of all the responses along the row corresponding to that hole depth. This arrangement can be expressed directly in matrix format, which for strain gauge calculations gives the lower triangular matrices $\bar{a}$ and $\bar{b}$ described in Chapter 5.

For strain gauge measurements, single-strain values are measured to identify each of the stress components P, Q, and T. Thus, a single number represents each loading case in Figure 5.8 within the matrices $\bar{a}$ and $\bar{b}$. However, when making full-field optical measurements, the number of measured displacement values for each loading case is in the 10^4–10^6 range. Under these circumstances, each number in matrices $\bar{a}$ and $\bar{b}$ is replaced by the much larger grouping shown in Figure 7.13 (or a reduced version for single-axis measurements) within a much larger combined G matrix. Figure 7.14 schematically illustrates the resulting matrix equation, where the rectangles represent entire matrices and vectors from Figure 7.13, each corresponding to its associated load case in Figure 5.8.

The G matrix in Figure 7.14 is very large, so significant care is required to handle it effectively. Fortunately, the corresponding $G^T G$ matrix for Equation (7.2) is much smaller, with dimensions $9n \times 9n$ (or $6n \times 6n$ for a single-axis measurement), where n is the number of hole depth increments used. This is a much more tractable size and is comfortably handled by modern computers. The various columns in the G matrix and δ vector, respectively, represent the theoretical and measured full-field images for the various load cases. These do not all need to be held in computer memory at the same time, thus avoiding the need to store the entire G matrix simultaneously. Instead, the various images can be loaded a few at a time and the $G^T G$ matrix progressively assembled by incrementing the dot products of corresponding parts of columns. Of note is that the parts of the G matrix corresponding to the artifacts w_4–w_9 are the same at each hole depth, so one copy of these numbers suffices for all hole depths. On a conventional 2017 desk computer, the computation time for a 16-increment single-axis measurement with 100,000 active pixels was approximately one minute. The required computation time varies approximately with the square of the number of hole depth increments and proportionally to the

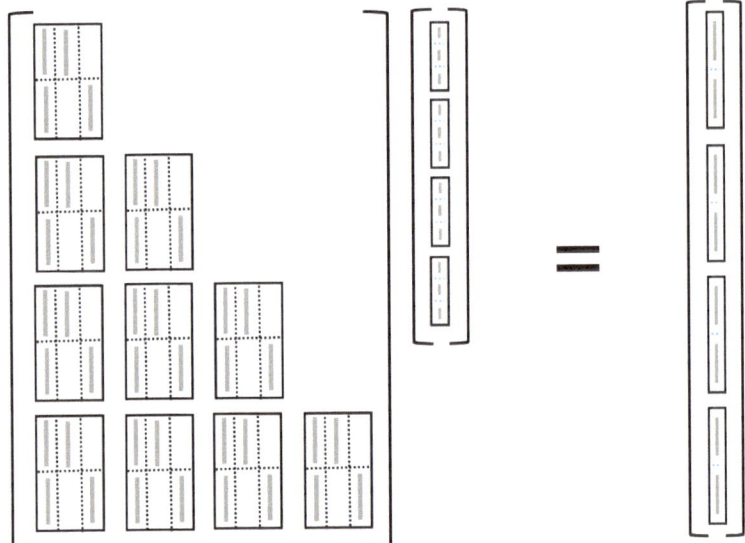

Figure 7.14: Schematic structure of Equation (7.1) for an Integral Method calculation.

number of active pixels. The majority of the computation time is used in assembling the $G^T G$ matrix because the G matrix in Figure 7.14 is so large. Solution of the resulting Equation (7.2) is fast by comparison because the assembled $G^T G$ matrix is symmetric and relatively modest in size.

 If desired, regularization can be added to the solution using a procedure analogous to that used for strain gauge measurements; see Equations (5.23) and (5.24). However, for optical measurements it is difficult to use the Morozov criterion to determine the amount of regularization needed because the least-squares solution gives a misfit (= residual) even without any regularization. In this case, it is practical just to adjust the amount of regularization until a reasonably but not excessively smooth solution is achieved. Fortunately, the data averaging that occurs during the least-squares solution of the optical data tends to reduce overall noise so that often little or no regularization is needed.

7.8   RESIDUAL STRESS COMPUTATION USING INCREMENTAL DATA

All measurement techniques for hole-drilling residual stress evaluations are differential in character. That is, they measure the difference between a starting state and a subsequent state. This is seen clearly when using strain gauges in that they must be "zeroed" after being installed and all subsequent measurements are made relative to that initial state. For hole-drilling measurements the starting state is the original uncut surface and the subsequent states occur after each increment of hole drilling. Optical measurements follow the same pattern, with an initial image made of the uncut surface followed by subsequent images made of the surface after each hole-drilling increment. The deformation of the imaged surface is then determined from the differences between the initial and subsequent images. This procedure works fairly well, but is not ideal for incremental measurements because the overall drilling process can take a significant time, during which time the optics may drift slightly, thereby causing a deterioration in the relationship between the later images and the initial image. This is particularly an issue with interferometry measurements such as ESPI and Moiré, where measurement noise tends to grow gradually with time from the initial measurement. Thus, in such measurements, it is an advantage for the initial image to be as "fresh" as possible.

The desire for a "fresh" initial image can be achieved during an incremental hole-drilling measurement by referencing each measured image to the one immediately before it rather than to the initially measured image. This practice significantly improves the quality of the measured data and substantially reduces the noise in the overall measurement. This is an important advantage given the noise sensitivity of the inverse solution required to determine the through-depth profile of the residual stresses.

Incremental image referencing can be accommodated mathematically with just a minor adjustment to the mathematical method described above. In Figure 7.14, starting from the second row of blocks in matrix G, the w_1, w_2 and w_3 contents of the blocks in the previous row are subtracted from each row to form a modified matrix G. This matrix can then be used in Equations (7.1) and (7.2), where δ takes the meaning of displacement during the previous hole depth increment only. To make this process clear, Table 7.2 reproduces the analogous, but much more compact strain gauge example $\bar{a}$ matrix from Chapter 5. Table 7.3 shows the equivalent $\bar{a}$ matrix for incremental strain referencing.

It was noted in Chapter 5 that the diminishing sensitivity of hole-drilling surface deformations to deep interior stresses causes the $\bar{a}$ matrix to be badly conditioned numerically, and thus to be very sensitive to measurement noise. This characteristic of the $\bar{a}$ matrix is revealed by the small numerical values of the elements along the diagonal relative to those away from the diagonal. However, inspection of Table 7.3 shows the opposite situation, with the diagonal elements within the incremental equivalent $\bar{a}$ matrix being larger then the off-diagonal elements. This makes the matrix much better conditioned numerically and therefore less sensitive to the effects of noise. The same effect occurs with matrix G in optical measurements, and so incremental referencing is advantageous, even with DIC, where temporal degradation of measured images is much less significant than with ESPI or Moiré measurements.

Figure 7.15 shows the results of an example incremental depth ESPI measurement on a shot-peened specimen. Figure 7.15a shows the stresses computed using optical data referenced from the initial measurement on the uncut surface, while Figure 7.15b shows the corresponding stresses computed using optical data referenced from the immediately previous image. It can be seen that the incremental results are much smoother and less prone to local noise disturbances.

Table 7.2: Matrix $\bar{a}$ for a 1/16" Type A rosette with a 2-mm diameter hole. Depths h and H are in mm. Multiply depths by 0.5 for a 1/32" rosette or by 2 for a 1/8" rosette.

h	[a]									
0.10	-0.0144									
0.20	-0.0189	-0.0152								
0.30	-0.0224	-0.0192	-0.0147							
0.40	-0.0250	-0.0222	-0.0182	-0.0132						
0.50	-0.0269	-0.0241	-0.0206	-0.0164	-0.0113					
0.60	-0.0285	-0.0256	-0.0221	-0.0184	-0.0141	-0.0093				
0.70	-0.0296	-0.0267	-0.0233	-0.0196	-0.0158	-0.0117	-0.0074			
0.80	-0.0305	-0.0276	-0.0241	-0.0205	-0.0168	-0.0131	-0.0093	-0.0056		
0.90	-0.0310	-0.0281	-0.0247	-0.0211	-0.0175	-0.0138	-0.0105	-0.0073	-0.0041	
1.00	-0.0315	-0.0286	-0.0252	-0.0216	-0.0179	-0.0144	-0.0111	-0.0082	-0.0054	-0.0028
H→	0.10	0.20	0.30	0.40	0.50	0.60	0.70	0.80	0.90	1.00

Table 7.3: Incremental matrix $\bar{a}$ for a 1/16" Type A rosette with a 2-mm diameter hole. Depths h and H are in mm. Multiply depths by 0.5 for a 1/32" rosette or by 2 for a 1/8" rosette.

h	[a]									
0.10	-0.0144									
0.20	-0.0045	-0.0152								
0.30	-0.0035	-0.0040	-0.0147							
0.40	-0.0026	-0.0030	-0.0035	-0.0132						
0.50	-0.0019	-0.0019	-0.0024	-0.0032	-0.0113					
0.60	-0.0016	-0.0015	-0.0015	-0.0020	-0.0028	-0.0093				
0.70	-0.0011	-0.0011	-0.0012	-0.0012	-0.0017	-0.0024	-0.0074			
0.80	-0.0009	-0.0009	-0.0008	-0.0009	-0.0010	-0.0014	-0.0019	-0.0056		
0.90	-0.0005	-0.0005	-0.0006	-0.0006	-0.0007	-0.0007	-0.0012	-0.0017	-0.0041	
1.00	-0.0005	-0.0005	-0.0005	-0.0005	-0.0004	-0.0006	-0.0006	-0.0009	-0.0013	-0.0028
H→	0.10	0.20	0.30	0.40	0.50	0.60	0.70	0.80	0.90	1.00

In addition, the x and y stresses are more consistently equal, which is a feature to be expected from a shot-peened specimen. These advances are due both to the improvement in quality of the incrementally referenced data and to the better mathematical conditioning of the incremental G matrix.

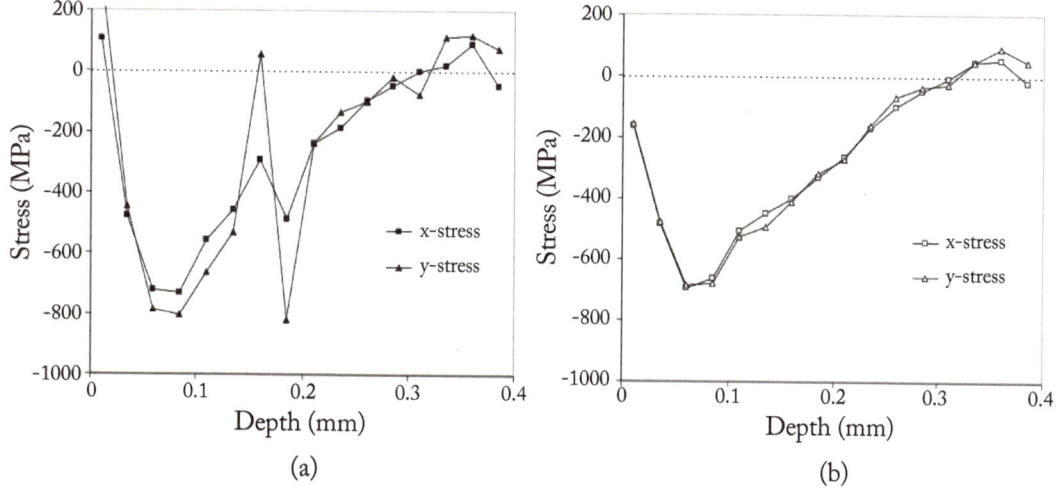

Figure 7.15: Hole-drilling ESPI measurements of residual stresses in a shot-peened material. (a) Computed using initial image reference and (b) computed using previous image reference (from Schajer and Rickert (2011)).

7.9   CONCLUDING REMARKS

Compared with the very well established strain gauge method, use of the various optical techniques is still very much in its infancy. At present, most reported optical measurements are research oriented; it is only recently that optical hole-drilling equipment suitable for routine measurements has become commercially available. A crystal ball would indeed be an appropriate device for predicting the future of optical measurement techniques. However, even in the absence of such an accessory, there is still a good basis for anticipating a continued strong interest in the development of optical techniques. Their rich full-field data certainly gives many fertile opportunities. For example, it seems very likely that optical measurements will provide a powerful tool to address the long-standing challenge of material plasticity when doing hole-drilling measurements in materials with residual stresses close to yield point.

Right now and for the foreseeable future, strain gauges may be expected to remain the standard choice for routine residual stress measurements. However, in an era when DIC capabilities are freely available as an iPhone app, it may reasonably be anticipated that the optical measurement techniques will grow to provide some serious competition.

7.10   FURTHER READING

Holography

- Jones R, Wykes C (1989). *Holographic and Speckle Interferometry*, 2nd ed., Cambridge University Press, Cambridge, UK.

- Lin ST, Hsieh CT, Hu CP (1994). Two Holographic Blind-hole Methods for Measuring Residual Stresses. *Experimental Mechanics*, 34(2):141–147.

- Nelson DV, McCrickerd JT (1986). Residual-stress Determination Through Combined Use of Holographic Interferometry and Blind-hole Drilling. *Experimental Mechanics*, 26(4):371–378.

- Makino A, Nelson D (1994). Residual Stress Determination by Single-axis Holographic Interferometry and Hole Drilling. Part I: Theory. *Experimental Mechanics*, 34(1):66–78.

- Nelson D, Fuchs E, Makino A, Williams D (1994). Residual Stress Determination by Single-axis Holographic Interferometry and Hole Drilling. Part II: Experiments. *Experimental Mechanics*, 34(1):79–88.

- Gabor D (1947). Apparatus for Producing Images of Small Objects by Photographic Means. Patent GB685286, Intellectual Property Office, London, UK.

ESPI

- Cloud GL (1995). *Optical Methods of Engineering Analysis*, Cambridge University Press, Cambridge, UK.

- Steinzig M, Ponslet E (2003). Residual Stress Measurement Using the Hole Drilling Method and Laser Speckle Interferometry: Part I. *Experimental Techniques*, 27(3):43–46.

- Focht G, Schiffner K (2003). Determination of Residual Stresses by an Optical Correlative Hole Drilling Method. *Experimental Mechanics*, 43(1):97–104.

- Viotti MR, Albertazzi A (2012). Compact Sensor Combining DSPI and the Hole-Drilling Technique to Measure Non-uniform Residual Stress Fields. *Proc. SPIE 8413, Speckle 2012: 5th International Conference on Speckle Metrology*, Vigo, Spain.

- Wu SY, Qin YW (1995). Determination of Residual Stresses Using Large Shearing Speckle Interferometry and the Hole Drilling Method. *Optics and Lasers in Engineering*, 23(4):233–244.

- Nelson DV (2010). Residual Stress Determination by Hole Drilling Combined with Optical Methods. *Experimental Mechanics*, 50(2):145–158.

Moiré

- Han B, Post D, and Ifju P (2001). Moiré Interferometry for Engineering Mechanics: Current Practices and Future Developments. *The Journal of Strain Analysis for Engineering Design*, 36(1):101–117.

- McDonach A, McKelvie J, MacKenzie P, Walker CA (1983). Improved Moiré Interferometry and Applications in Fracture Mechanics, Residual Stress and Damaged Composites. *Experimental Techniques*, 7(6):20–24.

- Nicoletto G (1991). Moiré Interferometry Determination of Residual Stresses in the Presence of Gradients. *Experimental Mechanics*, 31(3):252–256.

- Wu Z, Lu J, Han B (1998). Study of Residual Stress Distribution by a Combined Method of Moiré Interferometry and Incremental Hole Drilling. *Journal of Applied Mechanics*, 65(4) Part I:837–843, Part II:844–850.

- Ya M, Miao H, Zhang X, Lu J (2006). Determination of Residual Stress by Use of Phase Shifting Moiré Interferometry and Hole-Drilling Method. *Optics and Lasers in Engineering*, 44(1):68–79.

Digital Image Correlation

- McGinnis MJ, Pessiki S, Turker H (2005). Application of Three-dimensional Digital Image Correlation to the Core-drilling Method. *Experimental Mechanics*, 45(4):359–367.

- Nelson DV, Makino A, Schmidt T (2006). Residual Stress Determination Using Hole Drilling and 3D Image Correlation. *Experimental Mechanics*, 46(1):31–38.

- Schajer GS, Winiarski B, Withers PJ (2013). Hole-Drilling Residual Stress Measurement With Artifact Correction Using Full-field DIC. *Experimental Mechanics*, 53(2):255–265.

- Chan J (2015). iDIC Digital Image Correlation. iPhone app. https://itunes.apple.com/us/app/idic-digital-image-correlation/id995123721?mt=8

Computational Procedures

- Schajer GS, Steinzig M (2005). Full-field Calculation of Hole Drilling Residual Stresses from ESPI Data. *Experimental Mechanics*, 45(6):526–532.

- Baldi A (2005). A New Analytical Approach for Hole Drilling Residual Stress Analysis by Full Field Method. *Journal of Engineering Materials and Technology*, 127(2):165–169.

- Baldi A (2014). Residual Stress Measurement Using Hole Drilling and Integrated Digital Image Correlation Techniques. *Experimental Mechanics*, 54(3):379–391.

- Schajer GS (2007). Hole-Drilling Residual Stress Profiling with Automated Smoothing. *Journal of Engineering Materials and Technology*, 129(3):440–445.

- Schajer GS, Rickert TJ (2011). Incremental Computation Technique for Residual Stress Calculations Using the Integral Method. *Experimental Mechanics*, 51(7):1217–1222.

- Albertazzi A, Zanini F, Viotti M, Veiga C (2015). Residual Stresses Measurement by the Hole-Drilling Technique and DSPI Using the Integral Method with Displacement Coefficients. *Proc. of the 5th International Symposium on Experimental Mechanics and 9th Symposium on Optics in Industry (ISEM-SOI)*, 5:35–42, Guanajuato, Mexico.

- Furgiuele FM, Pagnotta L, Poggialini A (1991). Measuring Residual Stresses by Hole Drilling and Coherent Optics Techniques: A Numerical Calibration. *Journal of Engineering Materials and Technology*, 113(1):41–50.

Authors' Biographies

GARY S. SCHAJER

Gary S. Schajer is a Professor in the Department of Mechanical Engineering at the University of British Columbia, Vancouver, Canada. He has extensive experience with residual stress measurements, notably the hole-drilling method using strain gauges, Electronic Speckle Pattern Interferometry (ESPI), and Digital Image Correlation (DIC). Prof. Schajer has published extensively on residual stress and related topics and recently edited the reference book *Practical Residual Stress Measurement Methods*. He is a member of ASTM committee E28.13 on residual stress measurement and has been responsible for three major revisions of ASTM E837 Standard Test Method for Hole-Drilling Residual Stress Measurements.

PHILIP S. WHITEHEAD

Philip S. Whitehead is the Managing Director of Stresscraft Limited, located in Shepshed, Leicestershire, England. The company provides hole-drilling residual stress measurement services for academic and industrial customers, principally in the fields of aerospace and advanced material processing. Philip has 30 years of experience in the application of the strain gauge hole-drilling method and has designed and developed the many specialized hole-drilling devices used by Stresscraft. He continues to create novel drill head and strain gauge rosette designs to extend the application of the method. Current areas of interest include hole-drilling at remote features with limited access.

Index

Air turbine, 50
Air-abrasive, 52
Alignment lines, 73
Aluminium alloy block, 133
ASTM E837, 32, 55, 61, 69, 70, 87, 91, 95, 100, 105, 108, 111, 119
Autofrettage, 9
Average strain, 89
Average Stress Method, 60, 107

Bainite, 12
Bueckner's Principle,, 91
Bur (cutter) *see* Drilling cutter
Bur (rough edge), 73, 83

Calibration constants, 88
    dimensionless, 89, 91, 94
Casting, 6, 14
Characteristic length, 13
Chemical Vapor Deposition, 12
Coherency stresses, 13
Composite materials, 8
Continuous excitation, 76
Contour Method, 35
Crack Compliance Method *see* Slitting Method
Cumulative compliance function, 28
Cumulative strain function, 100, 101

Deep rolling, 11
Deep-Hole Method, 33, 49, 58

Destructive measurement methods, 16, 19
Deviatoric stress *see* Residual stress, deviatoric
DIC *see* Digital Image Correlation
Differential Strain Method, 60, 107
Diffractive methods, 20
Digital Image Correlation, 37, 56, 57, 150
Drilling cutter
    chamfered, 80
    coated, 80
    dental bur, 50, 80
    early designs, 51
    inverted cone, 80
    quality testing, 133
    tungsten carbide, 80
Dual-Trench Method *see* Two-Groove Method

E837 *see* ASTM E837
EDM *see* Electro-Discharge Machining
Eigenstrain, 3
Electro-Chemical Machining, 53
Electro-Discharge Machining, 11, 35, 53, 135
    re-cast layer, 136
Electronic Speckle Pattern Interferometry, 56, 148
Electroplating, 12
Equivalent Uniform Stress, 95, 108
ESPI *see* Electronic Speckle Pattern Interferometry

EUS *see* Equivalent Uniform Stress
Excision Method, 21
Extrusion, 10

Fatigue, 13
FIB *see* Focused Ion Beam
Finite element
    harmonic elements, 92
    mesh, 93
    method, 61
Fit-up stresses, 4
Flywheels, 9
Focused Ion Beam, 37, 57
Fringe pattern, 146, 147, 150, 153, 154
Full-field data, 143

Geological measurements, 58
Glass, heat-tempered, 1, 3
Grinding, 10

Heat tensioning, 12
Hole eccentricity, 110
Hole-drilling
    calibration constants, 100
    concentricity, 83
    depth accuracy, 127
    depth increments, 82, 95, 100
    distance to a step, 126
    distance to an edge, 125
    eccentric hole, 110
    equipment, 77
    geometric effect, 96, 108
    hole diameter, 122
    hole spacing, 126
    inspection of hole, 83
    maximum hole depth, 122
    measurement sequence, 127
    number of gauges, 127
    orbital drilling, 52, 78, 132

orthotropic materials, 112
procedure, 81
stress effect, 96, 102, 108
test data, 85
through-hole, 119
training sample, 132
Hole-Drilling Method, 31, 47
    concept, 47
    rosette selection, 70
    strain gauge technique, 69
Holographic Interferometry, 144
Honing, 11, 137

Incremental Slitting Method *see* Slitting
    Method
Ingot cracking, 7
Inherent strain, 3
Integral equation, 26, 102
Integral Method, 61, 98, 127, 159
Inverse equation, 26, 97
Isotropic stress *see* Residual stress, isotropic

Kernel function, 27, 34, 97
Kirsch solution, 60, 89, 109

Laser shock-peening, 11
Layer-Removal Method, 34
Least-Squares Method, 153, 156
Lens stresses, 4

Machined forged disc, 135
Magnetic methods, 20
Martensite, 12
Mathar, Josef, vii, viii, 47, 49, 58, 59, 63
Measurement method selection, 39, 40
Misfit, 2, 8, 106
Moiré interferometry, 146
Moiré Method, 56
Morozov criterion, 106, 160

Neutron diffraction, 20
Noise amplification, 104

Operator training, 133
Optical methods, 56, 62, 143
   features, 144
Orthotropic materials, 112

Phase change, 12
Phase stepping, 148
Phase unwrapping, 148
Photoelasticity, 1, 20
Plasma deposition, 12
Plastic deformation, 8, 10, 11, 14, 72, 83
Poisson's ratio, 89
Power Series Method, 108
Pre-stressed concrete, 14
Principal direction, 90
Principal stress, 90
Prong test, 24

Quenching, 6, 12

Rail neutral temperature, 8
Railway rail, 36
Regularization, 102, 160
Relaxation measurements, 16, 19, 20
Rendler and Vigness, 54, 55, 70, 91
Residual data, 154, 156
Residual stress
   description, 1
   deviatoric, 90, 97
   hidden character, 2, 15
   isotropic, 90, 97
   on unloading, 9
   origin, 2
   self-equilibration, 1
   strengthening effect, 9
   Type I, 13
   Type II, 13
   Type III, 13
Residual stress calculation, 59

   artifact compensation, 152, 157
   data averaging, 152
   data consistency checking, 152
   non-uniform stresses, 96, 158
   outlier correction, 152
   plasticity effects, 61, 93, 111, 163
   uniform stress, 87, 94, 152
   using incremental data, 160
Ring-Core Method, 29, 47, 49, 57, 88
Roller burnishing, 11
Rosette see Strain gauge rosette

Sachs' Method, 34
Scanning Electron Microscope, 37, 57, 152
Sectioning Method, 36
SEM see Scanning Electron Microscope
Semi-destructive measurement, 31
Shot-peened beam, 137
Shot-peening, 11, 14, 136, 162
Silver Bridge, 15
Slitting Method, 25
Smoothing, 104
Specimen
   curvature, 126
   geometry, 119
   intermediate thickness, 109, 137
   surface preparation, 71, 72, 73
   thickness, 108, 119, 120, 125
Speckle pattern, 148, 150
Splitting Method, 24
Spurious stress introduction, 72
St. Venant's principle, 29, 32, 96, 108
Stoney's Method, 12, 35
Strain gauge, 54, 69
   adhesive, 73
   backing material, 71, 129
   bond thickness, 129
   encapsulated grid, 71, 77, 120
   glueline, 71

installation, 73
instrumentation, 76
irregular surface, 129
non-standard rosette, 130
on poor conductors, 76
open grid, 71, 77, 120
practical installation, 122
shunt calibration, 77
thermal compensation, 70
thermal response., 82
three-wire connection, 77
witness marks, 71
Strain gauge rosette, 31, 54, 55, 69, 88
dimensions, 121
flexibility, 126
mean diameter, 119
selection, 119
size, 120
Type A, 55, 70, 94, 101, 119, 120
Type B, 55, 70, 119, 121, 128
Type C, 55, 70, 119, 121
Strain response function, 97
Strains p, q and t, 90, 97, 99
Stress averaging, 94, 108
Stress leveling, 10
Stresses P, Q and T, 89, 95, 97, 99
Structural redundancy, 5
Sun-kinks, 5
Surface machining, 10

Surface process samples, 135
Synchrotron diffraction, 20

Thermal expansion joint, 6
Thermal strain compensation *see* Strain
          gauge, thermal strain compensation
Thermal stresses, 5
Thermoplastic plate, 144
Thick specimen, 108
Thin specimen, 108
Tikhonov regularization, 105
Triangular interpolation, 101
Tube Splitting Method *see* Splitting Method
Two-Groove Method, 22

Ultrasonic method, 20

Vibratory stress relief, 6
Volterra equation, 26, 97

Welding, 6, 7
Wire drawing, 10
Wood, 15
Working conditions, 139

X-ray diffraction, 20

Yield stress, 32, 61, 122

Zero datum, 19